SpringerBriefs in Applied Sciences and Technology

PoliMI SpringerBriefs

W0079194

Series Editors

Barbara Pernici
Stefano Della Torre
Bianca M. Colosimo
Tiziano Faravelli
Roberto Paolucci
Silvia Piardi

For further volumes:
http://www.springer.com/series/11159
http://www.polimi.it

Marinella Ferrara · Murat Bengisu

Materials that Change Color

Smart Materials, Intelligent Design

POLITECNICO
DI MILANO

Marinella Ferrara
Department of Industrial Design
Politecnico di Milano
Milano
Italy

Murat Bengisu
Department of Industrial Design
Izmir University of Economics
Izmir
Turkey

ISSN 2282-2577
ISBN 978-3-319-00289-7
DOI 10.1007/978-3-319-00290-3
Springer Cham Heidelberg New York Dordrecht London

ISSN 2282-2585 (electronic)
ISBN 978-3-319-00290-3 (eBook)

Library of Congress Control Number: 2013941932

Printed on acid-free paper

Springer is part of Springer Science+Business Media (www.springer.com)

Contents

Chapter 1
Introduction

Abstract This chapter presents the subject and objectives of the book *Materials That Change Color* in the context of applied research and concept design. The implications of the techno-scientific research on theories and methods of design are outlined. Technical opportunities for new qualities given to objects of everyday life, like sensitivity, interactivity, communication skills and sustainability are analyzed from the perspective of design. In addition, the chapter discusses the implications of the extraordinary performance of the new materials on the broader meanings that new products present for users and on the visions of phenomenological reality in a cultural and socio-economic framework that characterizes the contemporary world. Some terms such as smart materials and smart systems are described and their role in design is shortly introduced. It is intended to provide designers the tools to develop skills necessary for innovative ways of thinking about materials and their relationship with technologies.

Keywords Smart materials · Smart systems · Design · Intelligent design · Nanoscale · Color

This book presents a design-driven investigation into *materials that change color*. These materials belong to the new class of high-performance materials, commonly known as *smart materials*, developed by chemists, physicists, materials engineers, and now available to be applied by designers to consumer products.

There is a vast variety of materials which make part of this new class of smart materials. In addition to materials that change color, treated in this book, there are those which change form, dimensions, temperature, those which transform one type of energy to another, and those which move; all in response to an external stimulus which induces a change in material properties according to the intrinsic nature of the materials, with a reversible effect.

This book introduces materials that change color with the aim of supplying basic information on them. Various categories of these materials are presented along with their behaviors in relation to the stimuli to which they react and other basic information, like characteristics, advantages, potentialities, production processes, and challenges for applications. This information will help to understand

M. Ferrara and M. Bengisu, *Materials that Change Color*, PoliMI SpringerBriefs, DOI: 10.1007/978-3-319-00290-3_1, © The Author(s) 2014

how materials that change color work, how they are applied to products and systems, and how their multi-faceted nature was put to use until today.

It is true that today's mode of techno-scientific operation allows the discovery, development, or replacement of smart materials continuously. Therefore, it is necessary to understand the properties of these materials and how they would behave under a certain energy input for new applications and for new directions in research in order to profit from them.

Another objective of this book is to develop a methodological approach for the use of these materials and related technologies, as well as a design vision, which is feasible and sustainable in the context of current problems. It is aimed to stimulate designers to take a more proactive attitude in the choice and application of these materials, in an efficient manner and as a strategy to realize innovations, which can contribute to social welfare and not merely aiming at commercial exploitation.

In order to aid these objectives, the book also presents a number of case studies: products, projects, concepts and experiments using smart materials, thus mapping out new design territories, roles and opportunities for these innovative materials. These case studies were chosen by the authors based on their capacity to represent state of the art projects and experiments in different fields of design, including product, interior, fashion and communication design. Unlike industrial patents, design case studies, by showing design methods and their results, are useful to understand both the functional and expressive nature of these materials. They show the new qualitative dimensions that smart materials bring into industrial and product design, the role that these new materials and technologies can play, and their influence in different areas of design.

The decision of treating a selection of case histories in a transverse manner in various fields of design has the intention of promoting a better understanding of opportunities offered by these technologies for designers. This approach derives from the typical mode of operation of Italian design, which utilizes tools such as design-driven innovation, cross-fertilization, and technology transfer, in order to develop creativity and facilitate innovation in products deriving from sectors with low capital investment. The whole of these case studies demonstrate the opportunities, which appear in projects with new scenarios for objects, environments, relations, and interactive systems, which satisfy new performance requirements. These case histories also form an ample range of scenarios of methodological approaches used by designers in various fields of smart material applications.

Within the context of rising sustainable and human-centered design agendas, this book will demonstrate the role and influence of these new materials and technologies on design, and discuss how they can implement and redefine our objects and spaces to encourage more resilient environments. The applications of smart materials are able to safeguard energy and material resources, to enrich product function, aesthetics, safety, and their communicative potential, and to contribute to a pleasurable user-product interaction. Once these potentials are put into action in a tangible project, it is possible to talk about *intelligent design*.

Smart materials provide new sources and perspectives for intelligent design. They are expected to be one of the key contributors to revolutionary products and systems of the near future. This book is meant to be a witness to the first steps toward a more human-centered, sustainable, and hopefully enjoyable future.

1.1 Smart Materials and Implications in Design

As it happens with all that is new, the widespread use of smart materials will also depend on their *acceptance* and *familiarity*. The acceptance may be facilitated through a comprehension, which is derived from correct information and from modes of communication, which must connect actors of innovation, namely materials scientists, engineers, designers, entrepreneurs, producers, distributors, and consumers/users. This is not an easy task but we want to try it at least for material scientists, engineers, and designers. This brief introduction intends to provide designers the tools to develop skills necessary for innovative ways of thinking about materials and their relationship with technologies.

Within the last five decades, materials became the true performers of change. One of the key innovations in that field has been the emergence of smart materials. This new class of high-performance materials is the highest expression of the *paradigm of tailor-made materials* (Manzini 1986; Ferrara 2004, 2005; Cardillo and Ferrara 2008) that has consolidated this contemporary third phase of the industrial revolution: the electronics and computer revolution. In other words, smart materials have been derived from the techno-scientific capability to intervene in the matter, not at the macro-scale, but at the molecular scale, and modify it[1] in relation to a project with predefined functions or performance. This model has given life to the electronic revolution, putting in action the potentials of artificial intelligence with an extremely reduced quantity of materials (miniaturization). As a matter of fact, working with materials at the electronic scale, the first transistor was realized, followed by all the electronic devices which we are using every day. Thus, the story of smart materials had started in the first 40 years of the 20th century when, thanks to the introduction of the electron microscope (1931), the study of materials at the electronic scale had amplified the understanding of the subatomic world, a dimension in which the laws of classical physics, where everything is measurable and predictable, was no more valid.[2] The study of

---

[1] In a program solicitation of the National Science Foundation it was stated as follows (NSF 2000): "One nanometer (one billionth of a meter) is a magical point on the dimensional scale. Nanostructures are at the confluence of the smallest of human-made devices and the largest molecules of living systems... A revolution has begun in science, engineering, and technology based on the ability to organize, characterize, and manipulate matter systematically at the nanoscale. Far-reaching outcomes for the Twenty-first century are envisioned in both scientific knowledge and a wide range of technologies in most industries, healthcare, conservation of materials and energy, biology, environment, and education".

materials has helped to understand their organization and operation, opening up the doors to the Quantum Theory and quantum mechanics.[3]

Unlike classical physics, which is based on the concept of solid and inde-structible particles (so-called *basic building blocks*), the concept of quanta, tiny energy packets with a double nature of wave (at subatomic level) and particle (at the moment of observation), made it possible to describe the dynamic properties of matter and the interaction of radiation with material.[4] It was thus concluded that the so-called building blocks are intangible energy waves, which appear like solid entity due to the great velocity at which they rotate.[5] A paradox of nature, which is

[2] All classical physics was constructed around the mechanistic Newtonian model of the universe in which all physical phenomena took place. This was the three dimensional space of the classical Euclidian geometry: an absolute space, always steady and unchangeable. All the changes occurring in the physical world were described as a function of a separate dimension, called time. Also this was absolute, which did not have any link to the material world, which was flowing evenly from the past to the future, through the present. The elements of the Newtonian world, which were moving in this space and in this absolute time were material particles. In mathematical equations, these were treated as material points and Newton considered them as small, solid, indestructible objects from which all material was comprised. This model was very similar to the atomistic model of the ancient Greeks. Both were based on the distinction between empty and full and between material and space. In both of the models, the particles remained identical to themselves in mass and form and thus matter was always conserved and essentially inert.

[3] The formulation of the quantum theory began when Max Planck discovered that the energy of thermal radiation is not emitted in a continuous manner but in energy packages. Einstein called these energy packages *quanta* and postulated that light and all other forms of electromagnetic radiation can present themselves not only as electromagnetic waves but also in the form of quanta. Light quanta, which gave the name to quantum mechanics, were subsequently been accepted as real particles and are now called photons. However, these are special particles lacking mass and always, in motion at the speed of light. At subatomic level, matter is not situated at precise locations but they demonstrate a *tendency to be present* in a certain place while atomic events do not occur with certainty at a determined time but they show a *tendency to take place*.

[4] Hence the acceptance of the foundations of quantum mechanics:

- The objective state of matter is characterized by a superposition of several states.
- There is no objective reality of matter but only a reality that is determined by *observations* of a person from time to time.
- The fundamental dynamics of the micro-world are characterized by contingency.
- It is possible that, under certain conditions, matter can "communicate at a distance" or could "appear" from nothing.

[5] The current concept of atom is that of a complex dynamic system with the dimensions at the range of one tenth of a nanometer, composed of various types of neutral and charged particles. Negatively charged particles, i.e. electrons, orbit around a central nucleus, circa hundred thousand times smaller than the atom, in which almost all the mass is enclosed. Only those electrons less attached to the nucleus participate in complex processes of activation (during which the atom literally changes form), which give place to the stabilization of the chemical bond between atoms of condensed matter.

difficult to explain if our senses are involved[6]: how can something exist which is both intangible and tangible at the same time?

Well, matter appears ambivalent. This could be explained as a set of discrete atoms and particles where each of which have their own role and individuality or as a whole a space, a field which acts through waves, applying a force at an indefinite point (Heisenberg et al. 2002). Yet, even in today's techno-scientific era, quantum theory is science, which is *put up with*, rather than being *accepted*. Not because its implications have little interest, but because they are so discomforting with respect to past certainties, as to be incomprehensible if not directly unacceptable.

Similarly smart materials put our certainties in crisis. In fact, according to common sense, material is substance; the substance from which things are made, perceived by the senses, are characterized fundamentally by mass and volume. Even the terms material and substance, due to the strong philosophical connotation, remind something which exists in itself, in a permanent and stable fashion. With smart materials, this idea is not valid anymore, starting from the relationship between material consistency and its appearance, between stable characteristics and their possible variation with time.

But what is intended by the smartness of a material? The smartness in a material (or a system) is determined by the relationship between its properties, its state, and the energy applied directly to the material. If this relationship influences the internal energy of the material, altering both the molecular/crystal structure as well as the microstructure, then the input will cause a change in material properties: the material absorbs energy that enters and undergoes a change. The change of internal structure is a common property of smart materials. If the mechanism modifies the state of energy of the material but doesn't affect the material itself, in that case, the reaction consists of an energy exchange from one form to another: the material remains the same but the energy undergoes a change. Such materials are typically not considered smart materials.

Smart materials are sometimes used as a critical part of a *smart system*. Such systems are typically composed of a *sensor* which has the function of sensing a change in the environment, a *control* group which processes this data to decide on the type of action, and an *actuator* which performs the desired action. Smart systems do not necessarily contain smart materials, so the ones that do contain them are properly called *smart material systems* (Smith 2005; Varadan et al. 2006).

In the field of design, smart materials have challenged the rationalistic theory that is based on the truth of materials (Dunne 2005) to substitute it with the slogan "sincerity and ambiguity" (Paris 2009). In fact, smart materials, unlike common ones, have two or more appearances according to the dynamic behavior that varies with time in response to fields of energy. This is an important distinction that undermines both the user and the designer and challenges the suitability of

---

[6] Up until 1982, scientists were not able to obtain an undistorted and direct image of atoms. In 1982, Binnig and Rohrer succeeded to get an atomic resolution image of the surface of silicon atoms with the scanning tunneling microscope (STM) that their team had developed (Wiesendanger 1994).

instruments used for design up to now (Addington and Schodek 2005). With smart materials, the objects and their immediate environments change, as do the ways in which they are conceptualized, tested, designed, and produced.

Another critical issue for design is the nanometer scale of some smart materials, which makes them very difficult to manipulate and process. Nanoscale features are difficult to analyze even with state of the art electron microscopes.[7]

Up until now, the application of smart materials at the nanoscale for the production of integrated components, miniaturized and incorporated into technical objects, increased the gap between the comprehension of the material and the function which it confers to electronic items. The gap between the electronic scale and the scale of objects (Dunne 2005) is primarily determined by the technical complexity, but also by the way of using smart materials. Manipulating materials at the electronic scale is even today a very difficult way for designers. This role is essentially delegated to scientists. In the field of materials for design, Manzini (1986) was the first one to explore the implications of design with innovative materials. He highlighted the opening of a new chapter in the history of design and the need to define a new framework, a vision, even a method, by which the designers could work with other actors of innovation to imagine a new interactive nature of products. As stated by Dunne (2005), Manzini has highlighted the potential of miniaturization, integration of multiple functions into a single object and aesthetic-decorative characterization deriving from new materials.

Many other important possibilities wait to be explored by designers. The potential presented by new materials to open new channels of communication between objects, electronic environments, and users need to be addressed. Furthermore, the advantage that smart materials offer in terms of energy savings and environmental sustainability are tremendously important today. Since most of the publications on smart materials are scientifically and technically oriented, the cultural, poetic, and practical aspects need much more exploration to do.

If scientists and engineers have been engaged in the development of these new materials for the past 20–30 years, designers have now the responsibility to find ways to develop applications, namely the appropriate adoption of technologies, to

---

[7] The invention of STM represents, even if only partially, the breakage of barriers between the atomic (or nanoscopic) world and the everyday experience of people. Despite the fact that the atomic structure was described by theoretical physics since the 1930s, even today, most of the experimental data is of an indirect nature, provided mainly by techniques such as spectroscopy, X-ray diffraction, and electron microscopy. However, tools such as STM and atomic force microscopy, which have the ability to see and manipulate single atoms, are becoming more common. This permitted to understand that at the scale ranging between a few nanometers and the dimension of a single atom (0.1 nm), the properties of materials depend strongly on the dimension. Thus, a metallic particle can become transparent, a semiconductor particle can change color, another one can melt at a temperature significantly lower than the common counterpart, etc. All this without changing the chemical composition but only acting on the size. The wonderful performance of many smart materials are linked to this effect.

ensure that they become available for the improvement of our daily lives. In other words, design is a powerful process of adopting a technology of cultural appropriation[8] (Mosse 2010) that involves understanding, acceptance of the implications and consequences, and that operates by manipulating it so that it can improve the lives of users.

Today there are many arguments to support the use of smart materials. With regard to the intrinsic sustainability of smart materials, it can be pointed out that the advancement of knowledge (due to scientific discoveries and technological inventions) with the subsequent evolution of culture, has given rise to new questions and changing needs of the scientific, creative and productive approach.

Smart materials have been specifically engineered to accomplish a particular performance objective thanks to their capacity to respond dynamically to the environment (Addington and Schodek 2005). In fact, they are characterized by their ability to detect and respond to stimuli from the environment (such as stress, temperature, moisture, pH, electric or magnetic fields), by a specific change of behavior, as for instance a color or shape or form change. They are also able to detect the intensity of the specific stimulus and respond accordingly (Jiang and Feng 2010). Each type of smart material acts as if to have a "genetic" code, coinciding with the performance design, providing a specific and reversible reaction. It processes actual behavior in an analogous manner to biological systems.

Moreover, smart materials are 'first law materials', which means they adhere to the principle of conservation of energy, that is, they can "change an input energy into another form to produce an output energy in accordance with the first law of thermodynamics" (Addington and Schodek 2005).

Smart materials can be produced at the miniaturized dimension, even at nanometer-scale, because of their relatively homogeneous nature. The possibility of micro- or nanoscale processing of these materials encourage the invisible integration within other materials, to achieve sensors, actuators, and MEMS embedded into objects and our environment, extending the limits of where computation can operate, and enabling materials to interact with their surroundings (Manzini 1986; Cardillo and Ferrara 2008; Coelho and Maes 2007). This integration also allows a reduction of components and quantities of materials.

The integration of smart materials gives extraordinary performance and new qualities to the objects of everyday life: sensitivity, interactivity, and communication skills are just some of the new qualities. But the use of smart materials in a certain application does not necessarily lead to intelligent design. The intelligence depends on the type of project and on the vision of the designer.

---

[8] Already in 1851, at the Great Exhibition in London, the architectural theorist Gottfried Semper affirmed the importance of design to "appropriate" the new tools that modern technology provides.

References

Addington M, Schodek D (2005) Smart materials and technologies for the architecture and design profession. Elsevier/Architectural Press, Amsterdam

Cardillo M, Ferrara M (2008) Materiali intelligenti, sensibili, interattivi. 02 materiali per il design. Lupetti editori di comunicazione, Milano

Coelho M, Maes P (2007) Responsive Materials in the design of adaptive objects and spaces. http://www.interactivespaces.net/data/uploads/185.pdf. Accessed 6 Apr 2013

Dunne A (2005) Electronic products, aesthetic experience and critical design. MIT Press, Cambridge

Ferrara M (2004) Materiali e innovazione nel design. Gangemi editore, Roma

Ferrara M (2005) Acciaio. 01 materiali per il design. Lupetti editori di comunicazione, Milano

Jiang L, Feng L (2010) Bioinspired intelligent nanostructured interfacial materials. World Scientific Publishing, Singapore and Chemical Industry Press, Beijing

Heisenberg W, Born M, Schrödinger E, Auger P (2002) Discussione sulla fisica moderna. Bollati Beringhieri, Torino

Manzini E (1986) La materia dell'invenzione. Arcadia, Milano

Mosse A (2010) Energy-harvesting and self-actuated textiles for the home: designing with new materials and technologies. DUCK J 2010:1–10

National Science Foundation (2000) Nanoscale science and engineering (NSE) program solicitation for FY2001 http://www.nsf.gov/pubs/2000/nsf00119/nsf00119.pdf. Accessed 6 Apr 2013

Paris T (2009) Sincerità e ambiguità dei materiali. In: Ferrara M, Lucibello S (eds) Design follows materials. Alinea, Firenze, pp 6–9

Smith RC (2005) Smart material systems: model developments. Society for Industrial and Applied Mathematics, Philadelphia

Wiesendanger R (1994) Scanning probe microscopy and spectroscopy: methods and applications. Cambridge University Press, Cambridge

Varadan VK, Vinoy KJ, Gopalakrishnan S (2006) Smart material systems and MEMS. Wiley, Chichester

Chapter 2
Materials that Change Color

Abstract This chapter introduces *materials that change color* with scientific definitions and explanations of different categories such as photochromic, thermochromic, and electrochromic materials. The behaviors in relation to the stimuli to which they react is presented and a general picture of the potential of these materials are given with some examples/applications of the most commonly used types. This information will help to understand how *materials that change color* work, how they are applied to products and systems, and how their multi-faceted nature was put to use until today.

Keywords Chemochromic · Chromogenic · Electrochromic · Mechanochromic · Photochromic · Thermochromic

Materials that change color are termed *chromogenic materials* and they are described as chameleonic because they reversibly change color as a response to changes in environmental condition (Hu 2010; Lampert 2004; Ritter 2007). In certain applications, a permanent color change is preferred, which is also possible with some chromogenic materials.

Another term which is used to represent the behavior of materials that change color due to an external stimulus is *chromotropism*. While the term *chromotropic* is used as a synonym to *chromogenic* (Fukuda 2007) we suggest that it is used in a similar sense to differentiate *thermochromism* from *thermotropism*, which is accepted to mean a reversible change in optical characteristics due to microstructural changes such as phase separation, change of particle size, aggregation, or isotropic-anisotropic phase transitions, in response to thermal changes (Seeboth et al. 2010; Ritter 2007). Thus, chromotropic materials can be regarded as a subcategory of chromogenic materials where *chromotropism* specifically refers to a change in optical characteristics (transparency or light diffraction) due to microstructural changes within the material that occur by the surrounding chemical or physical stimuli such as solvent, temperature, pressure, light or electrons.

The study of chromogenic materials starts with a comprehension of the natural laws expressed by the theory of color: the visible light which appears to us white is formed by various colors. The physical cause which produces the sensation of color is the interaction of luminous radiation with the electrons of the external substance or object we are looking at; the color that we observe is the light radiation reflected from surfaces, made of a complementary color to the absorbed radiation.

The technical principle by which these materials function can be explained by an alteration in the equilibrium of electrons caused by the stimulus, with a consequent modification of optical properties, such as reflectance, absorption, emission, or transmission. This process, named *chromism*, involves a change in the micro-structure or electronic state of substances. The majority of chromism in polymeric materials occurs in conjugated polymers. Conjugated polymers have loosely bound, freely moving electrons, similar to mobile electrons in metals. Chromic phenomena in such polymersare induced by various external stimuli which have the ability of altering the electronic configuration and energy transitions of electrons in the substance (Simon and Nilsson 2010; Durasevic et al. 2011).

Many of the natural compounds and now a number of artificial compounds of specifically characterized chromic properties have been synthesized.

Chromogenic materials are potentially promising for design-driven innovation. They are useful instruments for augmenting the functional, aesthetic, or communicative performance of objects and media, while saving energy with respect to traditional systems if suitably designed.

Modifying color and transparency, the most immediate among visible aspects of an object, these materials offer new opportunities for aesthetic demands and new possibilities for emotional involvement of users (Fig. 2.1). Thanks to these

Fig. 2.1 Squidarella, color changing umbrella by *SquidLondon* (founded by Viviane Jaeger and Emma-Jayne Parkes), 2011. From the exclusive *Paint Drip* Collection for Tate Modern Museum. When this umbrella is dry the color is *black*, *grey* and *white*, but once the umbrella is wet, the *multicolored drips* are revealed

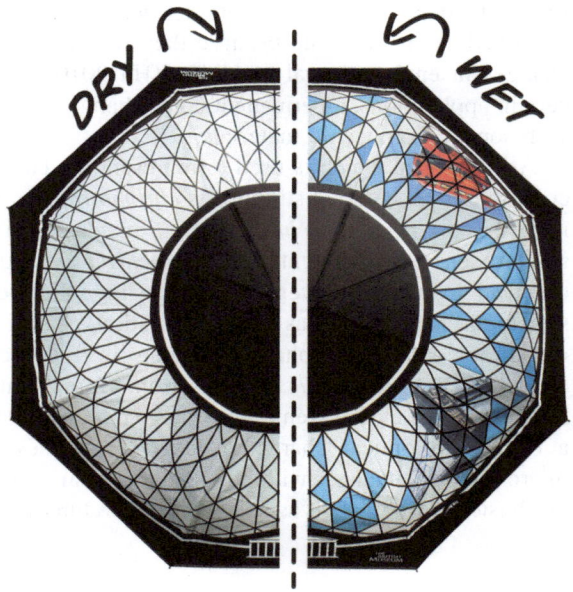

potentials, they are also used in the fields of art, such as in the work *Heart* by the Japanese artist Kiyoyuki Kikutake, exhibited at the Tokyo Museum of Modern Art.

In architecture, chromogenic materials opened new technological trajectories for the improvement of microclimatics and interior comfort, which reconcile energy savings with increased thermal and luminous performance. At the same time, they have contributed to the evolution of the architectonic languages for solutions equipped with dynamic qualities, characterized by reactivity and variability, which present new opportunities of interaction with users (Granqvist et al. 2010).

In product design, they improve the interaction between product and user, making the communicative interchange intelligent. Through the modification of color, it is possible to send messages and information to the users, such as communicating what is happening inside the product or how to use it. Chromogenic materials introduce new modalities to augment the reality of interactions, making it more continuous, persistent, and coherent to the feedback.

Because of all this, chromogenic materials offer a radical change in the ways of designing, perceiving, understanding objects and media, of interacting with the real world which surrounds us: a more immediate way, similar to the natural and biological logic and the human mind.

2.1 Classification of Chromogenic Materials

There are various categories among chromogenic materials, which take their names from the energy source, which provokes the modification of optical properties. It should be noted that the terminology given here is not common in all scientific disciplines. Depending on the specific field of study, many other terms have evolved which may have similar or overlapping meanings with the terms given here. For example the packaging literature uses the term *chemical indicator* in a similar sense to a *chemochromic* indicator. A temperature indicator for the packaging field is mostly described as a *thermochromic label* here. In the fields of biology and chemistry, *photobleaching* means the chemical destruction of a dye due to light (photons), which is more appropriately called a photochromic effect in the present context. Thus, we intend to use commonly accepted classifications mostly originating in chemistry and relevant to the fields of art, design, engineering, and materials science, but we do not imply that other terms are not acceptable. On the contrary, they may be more suitable for their specific fields.

• *Photochromic materials* materials in this category change their color when the intensity of incoming light changes. A well-known product which was developed using this principle is photochromic sunglasses. The lenses of these sunglasses get darker with increasing UV intensity and optimize the light that passes through them. When the UV intensity is lower, for example in the interior of a building, the lenses become more transparent and make it easier for the user to see through.

- *Thermochromic materials* respond to a variation in environmental temperature by changing their color. A well-known product which makes use of this phenomenon is a ceramic mug which changes color when a hot drink is poured inside. The transformation is reversible; thus the color of the mug goes back to its original one when it cools down to room temperature.
- *Mechanochromic/Piezochromic materials* show a change in color when a mechanical stimulus, i.e. stress is applied. These materials are intensely studied currently because of their potential use in stress detection, particularly for in situ failure monitoring due to fracture, corrosion, fatigue, or creep (Weder 2011).
- *Electrochromic materials* are characterized by an optical change upon the application of an electric field. A big market for electrochromic materials today is dynamic antiglare mirrors which detect glare and automatically compensate for it, especially for night time driving safety. Electrochromism is probably the most versatile of all chromogenic technologies because it is the easiest to control and it can be used in combination with different stimuli such as stress or temperature.
- *Chemochromic materials* respond to chemical changes in the environment by changing color. This phenomenon is used, for instance, to develop double pane windows with the ability to change color upon contact with hydrogen gas in the gap between the two panes (*gaschromic materials*). The WO_3 coating with the help of a catalyst layer switches from colorless to dark blue and reduces the transmittivity of light (Lampert 2004). *Halochromic materials* are a subgroup of chemochromic materials which change color as a response to pH changes in the environment. *Ionochromic* is a term similar to *chemochromic*, indicating a reaction to the presence of ions in a medium by a color variation. *Hygro and hydrochromics* make up another subgroup which react to humidity or the presence of water (Fig. 2.1) due to a phenomenon defined as *solvatochromism*. The classical definition of *solvatochromism* is a change in the absorption spectra of chemical compounds caused by the surrounding medium. This effect has been used for over a century to study solute–solvent reactions and solvent polarity in the UV, visible, and near-infrared regions of the electromagnetic spectrum (Reichardt 1994). For practical purposes and in the visible spectrum, solvatochromism is a color shift caused by solute–solvent interactions and this effect is being investigated, for example, to understand the behavior of synthetic and natural dyes in various solvents (Rauf and Hisaindee 2013).
- *Magnetochromic materials* respond to variations in magnetic fields applied to a substance. They are currently at their preliminary research phase but interesting applications may follow, such as reflective color magnetic paper (Hu et al. 2010).
- *Biochromic materials* Such materials were developed to detect and report the presence of pathogens with a color shift. Potential applications of biochromic materials include colorimetric detection of pathogens against food poisoning or bioterrorism.
- Several other chromogenic materials which are sensitive to special stimulations such as radioactivity, electron beams, or infrared radiation are also being

investigated. We predict that such materials will be more significant for scientific and engineering applications but some consumer products may also develop such as radioactivity detectors for individual users.

2.2 Photochromic Materials

Photochromic materials respond to variations of incoming light intensity or the spectral distribution of light, modifying their color reversibly. Photochromics are generally unstable organic molecules; they are transparent and colorless when the light is soft and they don't absorb light. Due to the induction effect of electromagnetic radiation by energetically rich photons of the near UV electromagnetic spectra, they are activated and change their molecular configuration and their light transmission coefficient, thus the chromatic spectrum at the exit, manifesting a color and reduced transparency. When the bright stimulus is removed, the color disappears because the material returns to its original molecular configuration.

Many biological systems exhibit photochromism. For exemple, rhodopsin is a natural photochromic substance, present in the retina of the eye (Klán and Wirz 2009). Rhodopsin is a pigment which is activated by light, producing a nerve stimulus transmitted to the cortex to start the process of visual perception. A similar protein was discovered in the bacterium Halobacterium salinarum (at that time named Halobacterium halobium) in 1971, which is called bacteriorhodopsin (BR). Various physical effects of BR have been discovered, including energy conversion (photosynthesis), photochromism, and photoelectrism (Durasevic et al. 2011; Hampp 2000).

Many inorganic materials, such as copper, mercury, various metal oxides and some minerals also exhibit photochromism (Durasevic et al. 2011; Van Gemert 1999). Inorganic photochromics are more appropriate for coatings on metal, glass, and ceramic surfaces while organic molecules such as spiropyrans, spirooxazines and fulgides are suitable for use on textiles (Durasevic et al. 2011; Dürr and Bouas-Laurent 1990) where they get a wide palette of colors across the visible range of light.

Photochromic materials in the form of pigments can be mixed with conventional materials and used in combination with other common pigments in order to obtain paints and inks with effects that vary from one color to another.

Some devices are made up of photochromic substances placed between two layers of conventional materials having different energy absorption. These materials, based on the absorption of light, are usually employed in the context where the limitation of ultraviolet (UV) and infrared (IR) radiation is the principal objective, such as in the case of optical lenses for solar protection, which represents one of the principal applications.

In the nineteenth century the phenomenon of photochromism has been studied to develop photochromic glasses. Now commercially available, those glasses can vary their coefficient of transmission as a function of light intensity. Through the

Fig. 2.2 Vision across a photochromic glass: on the *left*, in low light conditions the glass is transparent; on the *right*, in high luminosity conditions the glass is obscured. *Courtesy* Rudy Project

effect of silver halides and copper dopants contained in them, these glasses usually turn into a gray color when exposed to sunlight. In this mode, they function as a filter especially for infrared radiation, screening the heat emitted by the sun (Fig. 2.2). The gray color is a result of silver cations from the silver halide, combining with copper $(1+)$ electrons with the help of sunlight, forming metallic silver clusters. Photochromic glasses are also used for photographic reproduction and to realize "sensitive" frames and buildings. The use of photochromic windows has been limited until today due to certain technical and process-related problems such as obtaining a uniform dispersion of photochromic substances in the glass and the gradual loss of reversibility with time, also called fatigue. In some cases, experimental studies have already solved most of these technical difficulties which will allow the increase of photochromic sheet dimension and stability with time, reducing production costs and increasing the number of cycles or useful life of the products, thus making this technology a feasible one.

Among the commercial applications of photochromic materials are inks, which change color under the influence of intense UV light emitted by the sun, flashlight, and Wood lamps.[1] Very fine layers of the ink, deposited on a substrate, for example fabric or paper, or inserted between two polymeric films, change rapidly from an invisible state to an intense color of black, blue, or purple, maintaining the acquired tonality for approximately thirty-seconds. After the UV light is removed.

---

[1] Wood's lamps are used to emphasize fluorescent substances under UV-light. They have a wide range of applications in sectors such as mineralogy, archeology, restoration, biology, and for the verification of forgery in banknotes.

Furthermore, photochromic formulations of standard colors are possible, where for example yellow changes to red and red into purple.

Photochromic inks are applied on substrates via processes such as flexography, serigraphy, dry offset and typography (see Sect. 3.6). It is possible to formulate inks, which combine photochromic pigments with thermochromic or other chromogenic pigments. In order to obtain the best results, the photochromic ink layer should be as compact as possible and in the range of several micrometers. A rapidly growing field of application is in the sector of security inks used to print invisible images and text, which can only be revealed under UV light with the aid of suitable instruments.

In the textile sector, photochromic materials have been used for several decades now. In 1989, the first photochromic T-shirts were introduced to the market.

Currently, the fashion industry shows a good deal of interest in photochromic textiles, used to create colorful designs and effects, which change according to the incoming light intensity. The characteristics of these textiles depend on the substance used on them. The selection of the reflected waves depends on the type of molecules or liquid crystals used in microcapsules, which react in different ways to light.

The most applied synthetic substances used as dyes in the textile fibers are spiropyrans, spironaphtoxazines and chromenes. These compounds have been studied and show good resistance toward fatigue and photodegradation. Other compounds are fulgides and diarylethenes (Durasevic et al. 2011).

Methods of application are various. Some of these methods involve embedding the photochromic dye in the polymer matrix during the spinning phase of the synthetic fibers such as polypropylene. Dyeing and screen printing processes are considered more appropriate than competing processes since the demand on photochromic textiles is limited to piece garments, rather than to batch production (Billah 2008; Canal 2008; Durasevic et al. 2011; Nelson 2002).

Photochromic textiles have been used to tailor bathing suits or for work wear in low light conditions for increased visibility of the wearer. Photochromic dye onto textile material, if appropriately applied, forms a photochromic system, which is as a matter of fact a sensor capable of detecting and reacting to UV light of precisely defined spectra and intensity in a programmed and controllable way (Billah 2008). This kind of smart textile can alert and protect against the negative influence of UV irradiation, showing the color at the moment in which such irradiation occurs. This way, the increased functionality of photochromic textiles compared to conventional ones can be regarded as an added value in the textile product. On the other hand, the use of photochromic colorants in the printed form, which does not result in wastewater, is more convenient both economically and ecologically in comparison to conventional dyeing processes.

Photochromic systems have also been used for military applications, for example in optical memory devices (photosensitive glass to register messages). In biology and chemistry, they have been used to create molecular switches (Tian et al. 2003) and for optical data archiving in 3D (Corredor et al. 2006).

2.3 Thermochromic Materials

There are various categories among chromogenic materials, which take their names from the energy source, which provokes the modification of optical properties: Thermochromic materials change color upon a change in temperature. Mostly the color change is reversible, but there are also materials, which demonstrate irreversible color change (Fig. 2.3).

Materials belonging to different groups such as polymers, solid state semiconductors, or liquid crystals can show thermochromic behavior. The thermochromic effect can be abrupt at a certain temperature, or gradual, within a temperature range, depending on the material involved. This point (or range) is called the thermochromic switching or transition temperature. It is possible to adjust the switching temperature, T_s, by different approaches, depending on the type of material utilized.[2]

Fig. 2.3 A drop of Alsa thermochromic ink reacts to the heat generated by a match. *Courtesy* of Alsa Corporation

---

[2] Metal oxide thermochromic materials such as V_2O_3, V_2O_5, and Ti_2O_3 are semiconductors below a critical temperature, T_c, but behave as metals above T_c. In such systems, $T_s = T_c$ can be adjusted by doping. For example, vanadium (IV) oxide, VO_2, is the transition metal oxide with the closest T_c to room temperature ($T_c = 68$ °C). However this temperature is still too high for many applications. Doping VO_2 using metal ions with larger atomic radii than the V^{4+} ion reduces the switching temperature. Tungsten (VI), niobium (V), and titanium (IV) ions are such dopants. For example a 2 atm% addition of tungsten (VI) can reduce T_c to 25 °C. Similarly, it is possible to increase T_c by employing dopants with atomic radii smaller than the V^{4+} ion.

The capacity of these materials to adopt different color states at different temperatures and to change color and return to their original color countless times in response to temperature fluctuations makes them particularly interesting. The thermochromic effect is based on a chemical equilibrium between two different forms of a molecule or between different crystalline phases. Two possible types of phase transformation causing *thermochromism* are termed first order and second order, depending on their thermodynamics.[3] The color of transition metal oxides above the switching temperature can be affected by many factors. For example, heating of the metal oxide can change the oxidation state and this might result in errors of color interpretation because it may not behave as expected from a certain type of oxide. Purity and material thickness are two other important factors to consider. Below the switching temperature of 68 °C, VO_2 single crystals are transparent both to visible and infrared light. The monoclinic crystal structure transforms to a tetragonal structure at T_c and the material becomes metallic. While still transparent in the visible spectrum, it is now reflective in the infrared region. This effect is useful for adaptive solar control applications (Kanu and Binions 2010). When a thermochromic glazing reaches the switching temperature, it becomes reflective and prevents UV-radiation to pass through the window, controlling the amount of heat caused by sunlight. When the temperature decreases below T_c, for example in the afternoon or in cooler months of the year, the window becomes transparent to UV-radiation and allows making use of solar heating.

Among inorganic systems, the thermochromism of mercury iodide (HgI_2) is notable. This compound displays a yellow-orange color at ambient temperature but turns red when heated to 200 °C. The color switching is due to a crystallographic transformation (a change in the position of iodine ions around mercury ions) at the switching temperature, similar to thermochromic metal oxides.

Thermochromic colorants (pigments or dyes)[4] are produced and commercialized by many chemical companies for applications in various sectors. Such colorants are used to produce paints and inks for superficial finishing or they are mixed with other materials such as polymers for bulk coloration.

In the aerospace industry thermochromic paints are used to change the emissivity of surfaces under a thermal effect.

(Footnote 2 continued)
Aluminum (III) and chromium (III) can be used for this effect (Kiri et al. 2010). An alternative to doping is to strain thermochromic films or crystals to reduce the transformation temperature (Gu et al. 2010).

[3] V_2O_3, VO_2, and V_2O_5 undergo first order transformation, while Ti_2O_3 undergoes a second order phase transformation. These two types of transformations manifest themselves as abrupt and continuous transformations, respectively. Thus, thermochromic switching occurs suddenly at a given temperature for V_2O_3 (−123 °C), VO_2 (68 °C), and V_2O_5 (257 °C), while for Ti_2O_3 it occurs gradually within a broad temperature range (127–377 °C).

[4] *Dyes* are substances soluble in water, which impart their chromatic characteristics to materials onto which they are applied through processes of inclusion or chemical reaction. *Pigments* are fine powders, insoluble or little soluble in water, which need to be dispersed in binders, which fix them to the substrate.

Fig. 2.4 Jürgen Mayer H., Guest Book. This is a limited edition book composed solely of sheets printed with data protection pattern. A touchstone of the studio's work and focus of the exhibition *Patterns of Speculation*: J. Mayer data protection patterns are ubiquitous, gibberish-like image types that serve to conceal other information, including bank statements, shipping labels and paycheck stubs. *Courtesy* J. Mayer H. archive, Berlin

Inks can be used in suitable formulations on almost any type of substrate (paper, acrylic sheets—Fig. 2.4, textiles, metal, wood, etc.) through printing processes such as flexography, rotogravure, typography, offset, or screen printing. The availability of thermochromic paints and inks on the market and the case of application on various substrates has made it possible to experiment with them in the field of design.

Paints and inks have been used to augment and stress the visual appeal of certain potential products, with applications of various types. In fashion design, heat produced by the human can be used to activate the color change. In other projects and sectors, other types of heat sources have been used; for example dishes, pots, and pans can display a desired color when heated (Fig. 2.5); or some make use of the temperature change in heating devices such as in the case of the wallpaper project of Shi Yan (Fig. 2.6). In the latter two examples, thermochromic technology has been used to augment the performance of products.

In the case of pots and pans, the appearance or the color change warns us of the temperatures attained at the surface; in the case of the wallpaper, it informs us about the thermal energy which circulates the interior, translating the heat

---

[4] *Dyes* are substances soluble in water, which impart their chromatic characteristics to materials onto which they are applied through processes of inclusion or chemical reaction. *Pigments* are fine powders, insoluble or little soluble in water, which need to be dispersed in binders, which fix them to the substrate.

Fig. 2.5 One kettle by
Vessel Ideation. The kettle
body is of white enamel-
coated stainless steel, treated
with heat- sensitive inks. The
neck/grip area is covered with
high temperature silicone.
The kettle is normally white.
On the stove, after boiling,
the kettle shows a new face
ready to serve (courtesy
Vessel)

Fig. 2.6 Shi Yuan, thermochromic wallpaper, 2011. The flowered pattern only appears with the heat from the radiator (courtesy Shi Yuan)

sensation into a vision of vegetation that blooms in a tropical climate. In other instances the color change can be activated through resistive heating that employs a layer of conductive and thermochromic ink, which has been employed, for example, on battery testers.

Among thermochromic devices available in the market, there are those using liquid crystals as temperature sensors and fabrics used as actuators or interrupters, with the possibility of activating other devices without the need for electricity and without changing form.

The change of color is used to indicate temperature variations, such as in the case of plastic strip thermometers, in food packaging, in medical thermography

Fig. 2.7 Thermochromic Clock, 2011. It is a 4-digit 7-segment timepiece. You could say that it looks like any other digital timepiece, but this one just works quite different. Each segment in the display is made with a length of nichrome wire and then covered by a thick layer of black thermochromic (courtesy CW&T Art and Design studio)

applications, in engineering devices for non-destructive tests, and in electronic circuits. There are also many other cases where the application is purely experimental (Fig. 2.7).

In the textile industry, products dyed or coated with thermochromic materials are becoming more widespread thanks the thermochromic pigment powder diffusion (Fig. 2.8). Commonly two types of thermochromic dyes are being used on textiles: liquid crystal type and the molecular rearrangement type. In both cases, the dyes are entrapped in microcapsules and applied to garment fabric like a pigment in a resin binder. They have been applied since the 1980s on T-shirts and caps, which change color when the body temperature is increased.

In 1987 Toray Industries reported the development of a temperature sensitive fabric thanks to glass microcapsules, 3-4 microns in diameter, containing thermochromic colorants (the dyestuff, the chromophore agent acting as electron acceptor and the color neutralizer such as alcohol) coated homogeneously onto the textile surface (see Chap. 3) The product with the trade name Sway is a multicolor fabric, with four basic and 64 combined colors, which has the ability to change color reversibly at a temperature above 5 °C and operable between –40 and 80 °C. The change of color in these fabrics is designed to adapt to precise applications. For example, temperature ranges for winter clothing need to be between 11 and 19 °C, for women's wear between 13 and 22 °C, and for temperature shades between 24 and 32 °C (Project TeTRInno SmarTex 2007).

In textile applications, the process of coloration follows different approaches compared to classical colorants, as explained in Sect. 3.7. The most common thermochromic colorants used for textiles are leuco dyes and liquid crystals (Sect. 2.4.4).

Fig. 2.8 Thermochromic
pigment powder for use in
various applications

Liquid crystals allow to display temperature changes, because the change in color can be engineered and set at the decided temparature with tight tolerances. A common product where this precision has been employed is the liquid crystal thermometer, registering variations of 1 °C. Furthermore, some types of liquid crystals have an important advantage compared to leuco dyes. They can register the change in temperature passing through a series of colors; in other words they have the capacity of exhibiting different colors at different temperatures. However, they are delicate, usually require encapsulation, and they are typically more expensive than leuco dyes. Leuco dyes are more practical to work with, are more affordable, but their sensitivity is low, making them suitable for non-precision applications. They can be used at temperatures ranging from −25 to 65 °C and their color can be tuned to display a desired hue at a given temperature. The color can switch from clear to colored, or they can be mixed with permanent colored inks to switch from one color to another (Talvenmaa 2006).

Leuco dyes are mixed with an acid, which acts as an activator and a solvent, which transports the components. Thanks to the acid, physical transformations occur in the mixture which determine the switching from a colored aspect to a colorless one upon an increase in temperature, and vice versa upon cooling, according to the modality which depend on the chemical characteristics of the leuco dye utilized. By adding leuco dyes to traditional colorants it is possible to obtain unexpected color exchanges (Fig. 2.9). Mixing inks, which change color at different temperatures, one can obtain particularly interesting color effects (Figs. 2.10, 2.11, and 2.12).

Fig. 2.9 Linda Worbin, Textile Disobedience/Rather Boring (picture from her Phd thesis). This tablecloth is characterized by a reversible dynamic textile pattern made up of small X shapes in a perfect linear repetition, which reminisce the traditional cross-stitching technique. A certain part of the X's is printed with thermochromic pigments. When a hot air stream passes through the textile, these X's appear and reveal a hidden message

Fig. 2.10 Linda Worbin, Do Pattern (picture from her PhD thesis). The fabric is characterized by a matrix of polka dots printed in thermochromic pigment with dark color on a light background. Porcelain cups especially designed with different forms at the bottom interact with the polka dot pattern and create a dynamic pattern

For window applications, thermochromism has been made possible by the use of metallic oxide coatings such as vanadium dioxide (VO_2) or tungsten trioxide (WO_3). Another approach is to apply an intermediate gel layer between two layers of glass or plastic films.

Thermochromic coatings based on VO_2 offer optimum performance with regard to interior comfort based on an ideal relationship between switching temperature and environmental temperature. Gel systems typically depend on a considerable change in the light scattering properties at the switching temperature. This may be considered as a variant of thermochromism, which is termed *thermotropism*.

Thermotropism may occur due to different phenomena such as phase separation in a gel or phase transition between an isotropic and an anisotropic state.

One of the first significant trials of solar protection using thermotropic polymer gel was attempted at the residence of Munich Zoo between 1950 and 1960 with limited success (Seeboth et al. 2010).

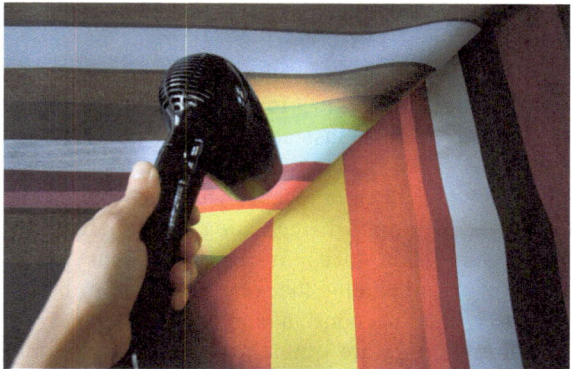

Fig. 2.11 Linda Worbin, Graffiti Cloth (picture from her PhD thesis). The textile combines three types of sensitive thermochromic pigments with conventional ones to explore the changes that occur when hot air flows. By using the *overprint* technique three more colors are added which become visible in the heated stage of the fabric. Some colors change from one to another while some colors stay in their initial state

Fig. 2.12 Linda Worbin,
Textile Disobedience
(picture from her PhD thesis).
A carbon fabric heating
element, connected to a 4,5 V
battery, is placed under the
textile printed in a *blue*
thermochromic textile.
Heating the fabric to a
temperature above 27 °C
creates a visible color change

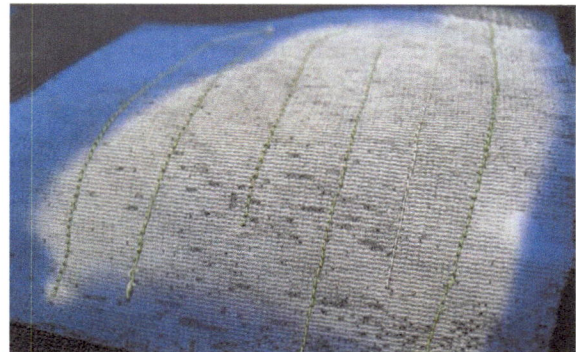

A thermotropic gel that has been commercially available since 1995 is Cloud Gel™. Cloud Gel is an aqueous polymeric solution, which is sandwiched between two external clear plastic films. The gel remains transparent up to a critical temperature, above which it appears white and the polymer chains that make up the gel curl up into balls, reflecting 90 % of the incoming solar radiation (Suntek 2012).

Applied in buildings, thermochromic windows allow to optimize the energy consumption of the building thanks to the high percentage of light transmitted both in the transparent as well as opaque state. Thermochromic windows automatically modify their transparency and thus their light absorption in relation to their external surface temperature. Initially transparent, above a critical temperature, which may vary from 10 to 90 °C, they become opaque. When the temperature decreases below the critical value, they return to being transparent.

The principal advantage consists of reduction of thermal loads due to seasonal heating and cooling, creating optimal climatic conditions and interior comfort. The light transmission, which characterizes these materials, is of diffused type, a specific quality which renders thermochromic materials suitable for applications where visibility is not desired but instead privacy or a screening effect must be guaranteed.

Compared to other types of chromogenic windows, thermochromic windows have some advantages. They are self-regulating with short response times; they reduce cooling and ventilation loads in an autonomous mode and eliminate problems of overheating by regulating the solar intake, contributing thus to energy savings; they diffuse the light in a constant and uniform manner both in the opaque as well as transparent state; they are simple to apply during building construction; they have low cost and long durability. They also have some characteristics which may constitute disadvantages: they are never completely transparent, thus they block beneficial solar rays in the winter; they can only be regulated using electrical circuits printed on the layers which cover the thermochromic film or layer; and some polymeric films have the tendency to turn yellow with time due to UV radiation (which can be solved by chemical stabilizers).

Industrial R&D is currently focused on certain aspects of thermochromic windows including control and modification of the switching temperature, control of luminous transition in the colored state, improvement of durability, and simplification of integration into architectural components.

Many applications of thermochromic materials have been realized so far. Some of them proved feasible and diffused to the global market. Thermochromic inks are used on beverage cans or bottles in order to indicate whether or not the optimum chilling temperature has been reached. Many brands are using this principle to indicate the consumer when the beverage is ready to be served. The ink can be directly applied to the metal can or on a label which is attached to the glass or plastic bottle. A common approach in the label or print design is to use a white area in the image which turns to blue when the product is chilled to the desired temperature (Hallcrest 2013).

Some applications were popular for a limited time. For example, thermochromic (so-called hypercolor) T-shirts created a boom effect in the '80s (Jones 2007) but did not last beyond the mid '90s.

Some other applications were tried on a smaller scale. For example, a thermochromic road sign has been used on Turkish highways (Fig. 2.13). The snow flake in the sign starts to change to blue below 2 °C and becomes completely blue below −1 °C in order to warn drivers about icy road conditions (Özgürler Trafik 2013). Although these road signs are still active, new ones are not implemented. Similar signs, for example in the UK, use small flashing lights instead of thermochromic devices, which are more effective in getting the attention of the driver. Nevertheless, other attempts exist to implement thermochromic paint for road signs (Nandu 2012), which may result in new products and applications in the future.

Fig. 2.13 Road sign showing risk of icing on a Turkish highway: "*blue color* indicates icing"—the snow flake turns to *blue* under −1 °C indicating icy road surface (photo by Murat Bengisu)

2.4 Electrochromic Materials

Electrochromic (EC) materials change color or opacity reversibly upon application of an electric field or transfer of an electric charge. Some scholars, as (Ritter 2007) underline the difference between electrochromic materials that are able to reversibly change color, and electro-optic (EO) materials that are able to reversibly change their optical characteristics (e.g. transparency or clarity), in response to electricity. Examples of EO materials would include polymer dispersed liquid crystals (Sect. 2.4.6), suspended particle devices (Sect. 2.4.1) and films incorporating micro-blinds (Sect. 2.4.1, Fig. 2.14).

The first discoveries on the phenomenon of electrochromism date back to 1953 when Kraus noted that upon application of an electric field, tungsten trioxide assumed an intense blue color. But it was Deb who from 1969 to 1973, upon studying the reaction of thin films of tungsten trioxide and molybdenum, defined the principles of electrochromism (Ritter 2007).

As these research studies demonstrated, the electrochromic effect occurs in certain inorganic materials including *transition metal oxides* (TMO) (oxides of tungsten, molybdenum, titanium, niobium, vanadium, iridium, cobalt, nickel).

Fig. 2.14 Film of micro-
blinds. Scanning electron
microscope image by
In-Hyouk Song of National
Research Council Canada
(through Wikimedia
Commons)

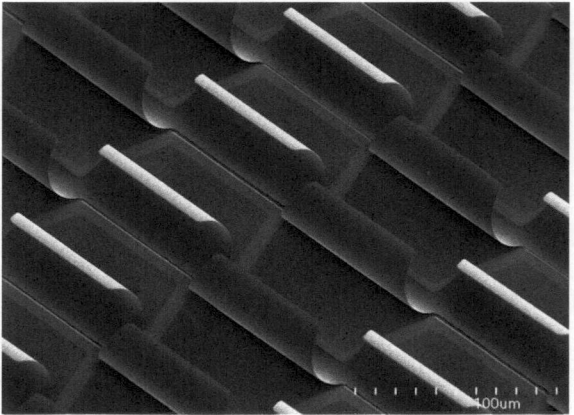

These oxides form the basis for devices which can be used in a large variety of applications ranging from sunglasses to intelligent windows, from antifog rear view mirrors to computer displays. Such devices consist of electrochemical materials whose chemically active electrodes are deposited in the form of thin films. Following an electrical input at relatively low voltage (1–5 V), the electrochromic process takes place in which the material undergoes a phase transformation and a change in color and transparency.

In the case of inorganic compounds, when electric current is applied, the electrochromic effect occurs by the mutual injection or extraction of positive ions (M^+) and electrons (e^-). The passage of electric current provokes a change of electron density in the compound in order to maintain electrical neutrality. This is the factor which modulates the optical behavior. In most of the solid metal oxide electrochromic materials, the atoms of these metals pass to a different valence from which they normally occur as oxides. These atoms at different valences, noted as *color centers*, are able to assign a certain color to a normally transparent structure.

One of the most commonly used electrochromic materials is tungsten oxide (WO_3) which occurs as a yellow, non-toxic, and solid powder at 20 °C. Its application temperature can range from <-40 °C to >120 °C. Used as cathodic coloring compound in electrochromic films, it reacts to the influence of an electrical field (by electrochemical reduction) and changes color from transparent to blue. It can sustain more than 100 cycles of color changes and shows a relatively good lightfastness compared to organic electrochromic compounds based on polymers (Ritter 2007). This material is the principal chemical substance used for the production of smart glass for windows. It can be coupled with reflective metallic surfaces of aluminum, silver, or nickel, to obtain optically active mirrors.

Another widely used inorganic material is indium tin oxide (ITO). In this case, the preparation of an electrochromic device implies the necessity to deposit many layers of materials onto transparent conducting electrodes, made of glass coated with a layer of ITO. Solid state devices based on this material contain distinct layers which are deposited successively onto a substrate such as a glass laminate and enclosed by a second protective layer to prevent the exposure to air and the consecutive damage of one of the terminal components. A first layer of highly conductive transparent film (transparent conductor, TC) is placed directly in contact with the substrate (solid substrate, SS). Afterwards, three layers are deposited: an active electrochromic material (electrochromic conductor), a layer of electrolyte of an ion conductive compound with high diffusivity (super ionic conductor) and a film, which acts as an ion reservoir (ion storage). The last one itself can also be made from an electrochromic material if the electrolyte possesses a high ion storage capacity. To guarantee the best adhesion between various layers and uniformity of surfaces, the assembly of the devic is realized in an autoclave at a pressure of about 10 atm. Finally, the composite is sealed and rendered hermetic for common applications.

Recent studies showed that the electrochromic effect also occurs in organic materials including various polymers (polyaniline, polypyrrole, polithiophene, polyisothiophene, and pyrazoline) (Fig. 2.15).

Most of these materials, both organic and inorganic, are electrochromic at the solid or viscous liquid state while so far no electrochromicity in the gas state has been demonstrated.

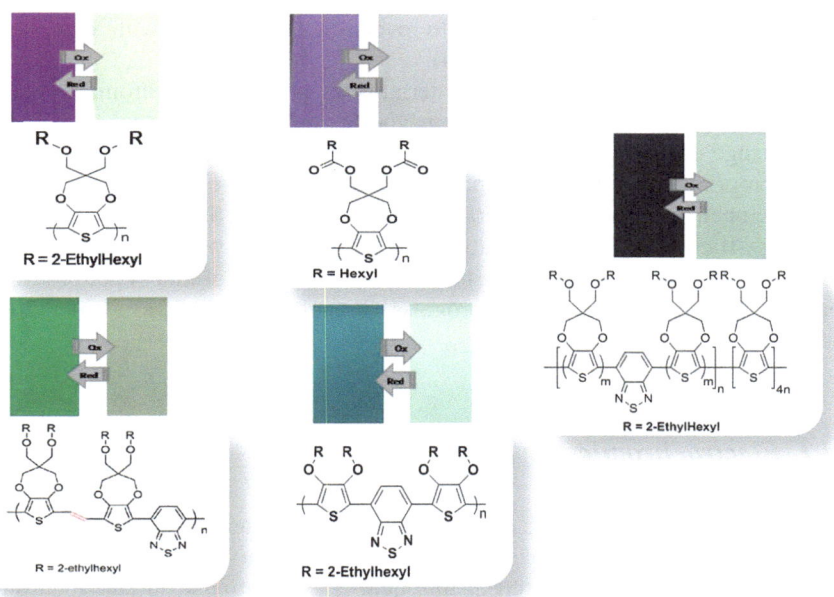

Fig. 2.15 Electrochromic soluble polymers on ITO-coated glass sides, in their neutral and doped states of color. Image of the Reynold Research Group Georgia Tech

An example to an organic electrochromic material is polyaniline, an electrically conductive polymer (produced by electrochemical polymerization of aniline), which is solid at 20 °C. Depending on the state of the electrochemical oxidation, caused by the presence of an electric field, the color of polyaniline changes from transparent to pale yellow and to dark green/black. Reversing the polarity reverses the color change.

Another group of organic electrochromic materials are polythiophene derivatives, also available in powder form, that change color from red to blue. Other colors are possible depending on the derivative.

Red and blue conducting polymers can be achieved by fine tuning just one absorption band to the desired value. It is more difficult to obtain other colors. In 2004, Fred Wudl and his group of researchers at the University of California developed an electrochromic polymer that appears green in its neutral form, thereby completing the red–green–blue color space and making any desired colour accessible by mixing. This polymer, a pyrazine derivative known as poly-DDTP, has two major absorption peaks, leaving a minimum at around 550 nm, which corresponds to the reflected green light. The researchers created polymer films and cycled them between the two differently colored states more than 10,000 times, without observing any changes in the spectroscopic properties, suggesting that the material is sufficiently robust for use in display technology (Sonmez et al. 2004).

A particular type of organic electrochromic materials are liquid crystals. This term defines substances which show a state of matter intermediate between a crystalline solid and an isotropic liquid, i.e., a solid–crystalline mesophase. This particular structure gives it the ability to reversibly change color with temperature and electrical stimulus (see Sect. 2.4.4).

A special type of electrochromic material is the photo-electrochromic, a hybrid combination based on dye-sensitized nano-materials. These types of materials which change color electrochemically but only on being illuminated, seem be the most appropriate candidates for smart windows. A photo-electrochromic system was developed in 2004, at the Fraunhofer Institute for Solar Energy Systems (Freiburg). It consists of a combination of a dye solar cell with an electrochromic component of tungsten oxide. The study and development of this system for large area window applications are in progress (Fraunhofer ISE 2013; Ritter 2007; Somani and Radhakrishnan 2002).

2.4.1 Electrochromic Devices

Electrochromic devices are one of the fields of smart materials applications with a huge commercial interest, thanks to their controllable transmission, absorption and reflectance. Most of the research in this field is concentrated on electrochromic windows of small- and large-area. Other possible applications such as flat panel displays, solar cells, frozen food monitoring, and document authentication are also of great interest. Given the vastness of the subject, we will limit our discussion to principal applications, research in progress, and general principles of operation, leaving the interested readers to explore the theme through the extensive list of references.

An electrochromic device (EC) conventionally consist in a "sandwich" configuration of electrodes, necessitating the use of at least one optically transparent electrode, such as indium tin oxide (ITO).

This type of EC, in it simpler configuration, is made of three layers of electrochromic materials deposited in the form of thin film on glass or polymer substrates. The middle layer is an ionic conductor (electrolyte); it loses ions when an electric current flows through it at relatively low voltage (1–5 V). The electrolyte layer is positioned between two layers: an electrochromic film, which acts as the electrode and a layer for the accumulation of electrons (counterelectrode) (Gillaspie et al. 2010).

A more complex scheme requires a structure with five layers. The first one is a transparent electron conductor deposited on glass or plastic, whose function is to secure an electric field uniformly distributed on the whole surface of the device. The second one is a layer of electrochromic material, which is able to conduct ions and electrons simultaneously. The third one is an ionic conductor or an electrolyte, whose function is to furnish or receive ions to insert to or to extract from the electrochromic layer. This layer conducts ions but not electrons. The fourth layer is the counterelectrode material, which is able to conduct electrons and ions simultaneously and to cede or store ions for the ionic conductor/electrolyte, needed for the functioning of the device. This layer can also be made from an electrochromic material. In that case, it must act in a complementary way to the second layer (coloration by anodic reaction if the other one is cathodic or vice versa). In this fashion, the two layers become colored together and they lose color together as well, assuring a higher modulation efficiency compared to a single electrochromic layer. The fifth layer is the second electron conductor, similar to the first one. Electrical connections are applied to this layer. In electrochromic mirrors and many display applications, this last layer does not need to be transparent since the device does not require it. In some cases, the fourth layer, in addition to being a reservoir of ions, is also an ionic conductor so efficient that it can substitute the third layer. Sometimes it also has electrochromic characteristics such that it can simultaneously substitute three layers (electrochromic, ionic conductor, and ion reservoir). In the case of devices based on organic semiliquid materials, two chemical species are mixed with a redox reaction on the surface of the electronic conductors, assuring optical modulation.

Already in the 1970s and 1980s, IBM and AT&T Bell Laboratories in USA, as well as Philips and Brown-Boveri in Europe, had experimented with the possibility of commercial applications of electrochromic materials, devices, and displays for use on a large scale. However, the prototypes degraded quickly over time which led to the rapid diffusion of the most dependable rival technology, namely liquid crystal devices.

Today, electrochromic technology has reached a higher reliability, and in the case of applications which do not necessitate a short response time, its performance characteristics may surpass those of common liquid crystal devices, providing, for example, better contrast and no limitation in the viewing angle.

Furthermore, compared to LCDs, in electrochromic displays, the electrical stimulus serves only to activate the process of coloration, which takes place gradually and progressively. This results in a smaller amount of electricity consumption during use. Today new types of electrochromic devices are available, making use of these advantages.

Liu and Coleman (2000) of Monsanto Corporation, developed a new fabrication method for flexible electrochromic displays without the use of transparent electrodes of conductive polymers. The method, defined *side-by-side*, use a simple printing technology and tin oxide nanocrystallites heavily doped with antimony (ATO) that exhibit a high level of electrochromism.

Recent nanotechnology advancements have enabled the creation of electrochromic displays using a thin film of about 30 micrometer, approximately one-third the thickness of a human hair. This is made of several stacked porous layers printed on top of each other on a substrate modified with a transparent conductor, such as ITO or PEDOT: PSS.[5] This device can be switched on by applying an electrical potential to the transparent conducting substrate relative to the conductive carbon layer. The electrical potential provokes a reduction of viologen molecules (coloration) to occur inside the working electrode. By reversing the electrical potential, the device bleaches. Furthermore, this electrochromic device needs a low switching voltage (around 1 V). This can be explained by the small over-potentials needed to drive the electrochemical reduction of the surface adsorbed viologens/chromogens (Somani and Radhakrishnan 2002).

Organic electrochromic materials based on the organic material PEDOT are currently considered to be the most suitable ones for applications on microfiber textile substrates.

Another new and interesting type of EC technology, promising for future development, is *micro-blinds*, able to control the amount of light passing through glass substrates in response to applied voltage.

Micro-blinds are composed of rolled thin metal blinds, practically invisible to the eye (Fig. 2.14). With no applied voltage, the micro-blinds are rolled and let light pass through. With a potential difference between the rolled metal layer and the transparent conductive layer, the electric field formed between the two electrodes causes the rolled micro-blinds to stretch out and thus block light. The metal layer is deposited by magnetron sputtering and patterned by laser or lithography process. The glass substrate includes a thin layer of a transparent conductive oxide (TCO) layer, a thin insulator deposited between the rolled metal layer and the TCO layer for electrical disconnection. Micro-blinds have several advantages: switching speed (in the order of milliseconds), UV durability, customized appearance and transmission. Theoretically the blinds are simple and cost-effective to fabricate.

---

[5] Poly(3,4-ethylenedioxythiophene) poly(styrenesulfonate).

A common technology for ECDs is *suspended particle devices* (SPDs). Also known as electrophoretic technology, this system consist of a thin film laminate where millions of small light-absorbing microscopic particles are distributed uniformly in an organic fluid and placed between the two panes of glass. Electrophoretic technology uses the up and down movement of bluish-black colored absorption particles that are suspended in a cross- linked polymer matrix to control light transmission when applying an electrical voltage signal. Upon application of an electric field, the particles align in the same orientation and let the light through. When the electric field is removed, the particles become dispersed in random orientations and absorb the light, so the film looks opaque dark, blue or grey or black (depending on the oxide applied). This type of technology is used in smart windows, displays and electronic paper (e-paper). In smart windows, SPDs can be manually or automatically "tuned" to precisely control the amount of light, glare and heat passing through, reducing the need for air conditioning during the summer months and heating during winter. Other advantages include reduction of buildings' carbon emissions and the elimination of a need for expensive window dressings.

Today, the market offers electrochromic devices of large or small surfaces for applications in various sectors (displays for watches, portable computers and other electronic products, information displays for advertising, big windows for building and automotive, etc.) even if one of the major advantages of electrochromic devices is their suitability for applications which require large surfaces (for example large windows in buildings). This advantage is lost with displays and applications of small size. Yet in this case, electrochromic materials represent certain advantages such as high contrast and non-restricted viewing angle. Furthermore when the power supply is interrupted, these materials retain their state for a long time. These benefits, however, should be confronted with their relative slowness in responding to electrical fields, degradative effects due to tens of millions of on/off operations that a display must endure during its lifetime, and the need to bring electrical connection to each pixel of the display.

Electrochromic displays still present technical problems because each pixel represents a single cell. This causes difficulties in the application of the multiplexing technique, which would permit the control of each row of the display pixel matrix, instead of the current need to control each individual pixel at a time.

Some electrochromic devices of small dimensions have been developed in the automotive sector, such as electrochromic sunroofs by Pilkington from UK and Flachglas AF from Germany (Lampert 1995). US companies OCLI/Donnelly and Gentex developed electrochromic rear-view mirrors and Gentex sold nearly 150 million of these products in the last 20 years (PPG Industries 2013). This device is equipped with two sensors placed on the external surface which determines the difference in brightness between the environment and the light coming from an object such as the headlights of the car which is in the back. The device automatically changes the intensity of reflection on the mirror, protecting the view from glare. These mirrors are now standard equipment in the higher end models of many automotive brands.

Other applications can be found in luxurious cars, such as Maybach or Ferrari Superamerica, designed by Pininfarina, whose sunroof is made by Saint Gobain Sekurit and launched in 2004. This sunroof in EC glass allows choosing the level of transparency inside the vehicle, thanks to a five-position selector which can be switched from the darkest color to the lightest in less than a minute.

2.4.2 Electronic Paper

Electronic paper (e-paper) is based on electronic ink (e-ink) and applied as display in e-book. The first example of e-paper was created at Xerox Parc in the 1970s by the researcher Nick Sheridon. In 1996 it was further developed by Joseph Jacobson, professor in the MIT Media Lab and one of the founder of E-ink Corporation. In 2004 Sony in collaboration with Royal Philips Electronic and E-ink Corp. realized the first e-book, the Librié EBR-1000, pocket-sized, capable of containing the digital memory of a whole library and updating it continuously by connecting to the web. In 2005, during the IDC European IT Forum in Paris, Nicholas Negroponte introduced the tablet PC which applies the electrophoretic bistable effect of e-ink, developed since 1997 at MIT, Boston.

The e-book display allows one to visualize text and images with effects very similar to printed paper, with even better contrast. It differs from a normal display because it does not use backlighting and reflects the ambient light as would do a piece of paper. Its main advantage compared to a backlit screen is the ability to read even under direct light conditions (Fig. 2.16).

According to the most widely used technique, the e-book display is produced by a multilayer polymeric composite that incorporates spherical microcapsules greatly reduced in size and which contain electrochromic pigments that react to electrical stimuli at low voltage (Fig. 2.17).

Fig. 2.16 Image of Jane Austen's eye on the e-ink screen of the Kindle 3 under direct light conditions. Photo by Chris Gray through Flickr

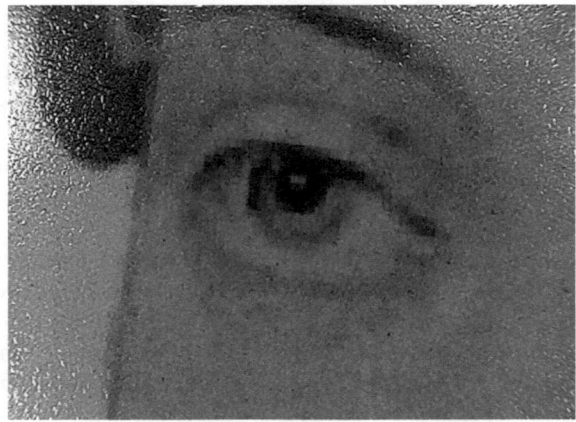

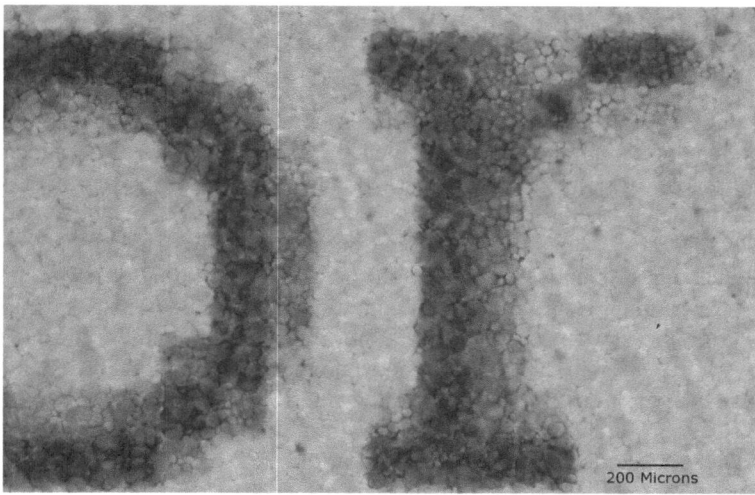

Fig. 2.17 View of an E-ink display under a microscope ($\sim$40X). The micrograph reveals the presence of microspheres of the size of several micrometers. Photo by Specius Reasons through Flickr

The spheres are electrically charged and have a positive part where the black pigments are located and a negative part where the white pigments are located. Through the electrical field, the spheres orient themselves to change the gray level in different parts of the screen (Fig. 2.18).

Recent studies helped the Chinese firm Hanvon to obtain e-papers capable of displaying various colors thanks to a layer placed on the micropheres which filters the light reflected from the spheres themselves.

In 2013, E-Ink introduced *Spectra* three colors electronic paper displays (white, red, and black) planned for electronic signage, announcing that other colors will be produced in the future.

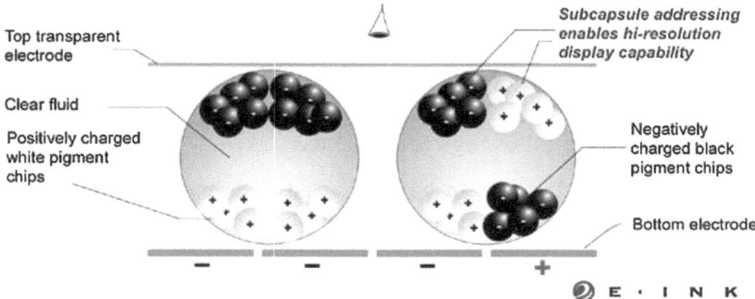

Fig. 2.18 *Cross section* of electronic-ink microcapsules by E-Ink Corporation

2.4.3 Electrochromic Devices for Windows

Over the past 20 years almost 1,000 patents and 1,500 papers in journals and proceedings have been published on *electrochromic windows* (Pawlicka 2009). The amount of research and development is increasing due to the advances in nanotechnology, which has a major impact on materials for electrochromic devices and on complete systems.

Today, the most promising technology for smart window is electrochromic. The EC glass changes its optical properties at the click of a button, or programmed to respond to changing sunlight and heat conditions in smart systems. Thus it allows control over the amount of solar radiation, light and heat passing through, providing visibility even in the darkened state and thus preserving visible contact with the outside environment. Electrochromic windows are useful in buildings (for windows and skylights), vehicles and aircraft, according to a rational energy management scheme and environmental comfort.

EC window multilayers are designed such that they allow the storage of ions and their movement back and forth for the insertion to or extraction from the electrochromic layer, through the application of an electric voltage. An electrochromic thin film stack with a thickness of about one micron is deposited on a glass substrate. The stack consists of ceramic metal oxide coating with three electrochromic layers sandwiched between two transparent electrical conductors (Granqvist 1995).

When a voltage is applied between the transparent electrical conductors, a distributed electrical field is set up. This field switches the glazing between a clear and transparent blue-gray translucent state, with no degradation in view, similar in appearance to photochromic sunglasses. The changing color can be modulated to intermediate states between clear and fully colored.

Furthermore, thanks to advances in transition metals, it was possible to develop hydride reflective electrochromic windows, which become reflective rather than absorbing, EC antiglare glass which satisfy requirements of color upon request and EC windows with low-emittance coatings and an insulating glass unit configuration which can be used to reduce heat transfer from this absorptive glazing layer to the interior.

The principal characteristic of electrochromic devices is the low energy consumption both during switching of the state as well as for storing the state of color in which the glass is located (this occurs without any energy consumption).

Baed on a comparison of data sheets of commercial products, ECDs for windows should satisfy the following requirements:

- Lifetime of at least 25 years in buildings and five years in automotive applications. *Solid state* devices are considered to be more durable (about 30 years) as opposed to *polymer laminate* devices;
- A total of about 5,000 cycles/year;
- Availability of color variations and optical modulation based on user requirements and changing environmental conditions;

- An on/off switching time of less than three minutes.
- Chemical-reactive stability.
- No toxic substance emission, not even in the case of fire.

Since the end of 2012, Sage Electrochromics Inc. (Minnesota), specialized in window glass development, produces a new electronically tintable glass (Fig. 2.19). It incorporates nanotechnology: five layers of ceramic materials, for a total thickness of less than 1/50th that of a human hair.

On hot and sunny days the glass would darken to reduce glare and block out heat; on cold, cloudy days the windows would clear to allow sunlight and heat to fill the interior.

The electrical switching of tints can be operated manually or integrated into an automated building management system. Applying a low voltage (less than 5 V) the glass darkens as lithium ions and associated electrons transfer from the counter electrode to an electrochromic electrode layer. Thus, sunlight and heat are absorbed and subsequently reradiated from the glass surface to the exterior—much the way low-emissivity glass also keeps out unwanted heat. Reversing the voltage polarity causes the ions and associated electrons to return to their original layer, clearing the counter electrode and the glass.

SageGlass blocks 91 % of the solar heat gain (Electrochromics Inc 2009). According to the U.S. Department of Energy's (DOE) Lawrence Berkeley National Laboratory, SageGlass technology "has the potential to reduce building

Fig. 2.19 Sage Electrochromic windows with tintable glass. Sage Electrochromic Inc

heating and air conditioning equipment size by up to 25 %, resulting in construction cost savings. SageGlass could also potentially reduce overall cooling loads for commercial buildings up to 20 % by lowering peak power demand and may reduce lighting costs by up to 60 % while providing building occupants with more natural daylight and greater comfort."

Electrochromic technology can also be used indoors to separate interior space and for protection of objects under glass from the damaging effects of UV and visible wavelengths of artificial light, for example in museum display cases and picture frame glass.

2.4.4 Liquid Crystals

The discovery of the liquid crystalline property possessed by certain organic compounds took place in the last decades of the nineteenth century (about 1888) following the studies of the Austrian botanist Reinitzer. While he was preparing cholesteryl benzoate, he realized that the substance seemed to present two distinct melting points with the formation of initially a rather cloudy liquid phase which successively turned perfectly clear at higher temperature (Larson 1999). He retained that the intermediate state was nothing but a new phase of the material where in conditions of typical fluidity of the liquid state, there was a certain molecular order, which resembled the order of solid crystals. Because of the intermediate characteristics between those of isotropic liquids and those of crystals, the new phase was properly called liquid crystal.

In a solid crystal, atoms (or molecules in the case of polymers) display a great degree of order both positional as well as orientational order, with very little freedom of displacement. In contrary, in an isotropic liquid, molecules do not have any particular order while they have ample freedom of movement. The molecules of a liquid crystal have characteristics in between, conserving a certain freedom of displacement whilst showing a tendency to assume preferential positions and orientations.

As it can be easily guessed, because of these qualities, liquid crystals (LCs) show unique properties, which have been the subject of numerous studies by researchers. Principally the anisotropy, which occurs due to preferential orientation of the liquid crystal molecules, makes them useful in many commercial applications; liquid crystal displays (LCDs) are just one example of these technologies.

The years following the studies of Reinitzer, a great deal of research has been dedicated to liquid crystals, which led to many new discoveries inventions. Primarily the differences between the liquid crystal phases have been determined thanks to studies by Demus and Richter (1979) and after the phases with positional order thanks to Gray and Goodby (1986). These researchers have generated big efforts to identify and characterize the liquid crystalline phases of newly synthesized compounds. It was observed that phases which have liquid crystalline

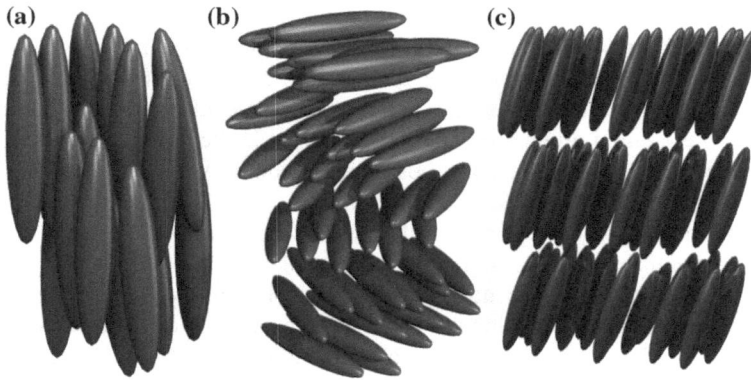

Fig. 2.20 Schematic of mesogen alignment in a liquid crystal nematic phase (*left*) and schematic of mesogen order in the cholesteric (chiral nematic) liquid crystal phase (*center*). The rotation between adjacent mesogens gives rise to an overall chiral phase. Schematic of mesogen order and alignment in liquid crystal smectic-c (*tilted, layered*) phase (*right*) (Image by Kebes trough Wikimedia Commons)

properties can be different in nature according to the level of molecular order and temperature. They were named *mesophases* while consecutively all organic compounds which are able to generate mesophases were called *mesogens*.

Mesophases (Fig. 2.20) are grouped according to their symmetries as:

- *Nematic*
- *Nematic-chiral* (or cholesteric), and
- *Smectic* (distinct in *A*, *B*, *C* and *D*, ferroelectric, antiferroelectric and *V*-shaped crystals).

Each of these phases has the peculiar characteristic that the molecules orient themselves preferentially along a common axis, i.e. the *director* (Gordon 2004).

The nematic mesophase, typically at higher temperature and liquid condition, is constituted of molecules packed without any positional order while they are characterized only by a common orientation, or the director. The more the average alignment of the molecules, the more accentuated the anisotropy of this material.

The cholesteric mesophase is tightly correlated with the nematic one. It is actually a variant of the nematic phase. In practice, the presence of intermolecular interactions causes a condition where each molecule remains disaligned a few degrees with respect to those which circumvent them, with the result that the orientation of the molecules is not constant along the whole phase. The director thus follows a helical path whose pitch is inversely proportional to the reciprocal disalignment of the molecules. The disalignment of the molecules can be increased by increasing the temperature, which tightens the pitch and decreases the pitch length, and vice versa. This property has a particular importance because cholesteric mesophases are able to selectively reflect visible light which has a wavelength equal to the length of the pitch. A variation in temperature thus manifests itself in a variation of the color of the liquid crystal.

The smectic mesophase (at lower temperature and solid condition) is characterized, with respect to the nematic, by a major molecular order. In addition to possess a preferential orientation, the molecules are in fact arranged in superposed planes with only a minor amount of positional freedom. Again, as it happens in the case of cholesteric mesophases, the director generates a helicoidal structure to which the macroscopic properties of the phase strongly depend.

The mechanism of optical switching in LCs changes the orientation or twist of LC molecules between two conductive electrodes, by means of an applied electric field. Reflectance and transmittance of the device (display or window) is directly related to the orientation of the LCs. The voltage required to switch LCs between bright and clear states is typically 24–100 V in PDLC devices and 3–15 V in LC displays.

Liquid crystals can be processed into films, to produce electro-optic layers suitable for several applications. One example in today's market is *Nanotec* film by Unilux (Fig. 2.21), a thin and flexible film with consists of a sandwich of two layers of PET that holds a polymer network ensuring a constant and super-fast conduction of electricity. This consists of a sponge-like structure that is filled with liquid crystals. In the *off* state, the material is characterized by a milky coloration due to the random alignment of nematic droplets, causing incoming light to diffuse. When a low voltage (48 V) is applied to the film, it changes from milky white to a transparent state. In fact, LC molecules orient themselves parallel to the electric field, so the incident light is not blocked or refracted and the material appears transparent. By adjusting the voltage, transparency can be adjusted.

Fig. 2.21 Nanotec film in LC by Unilux. The film is suitable for applying onto windows like a sun screen, onto objects, and is suitable for retrofitting. It can be used in a wet room, for example on shower doors and walls. It can be made into any size upon customer request. The film is currently available in white or black color

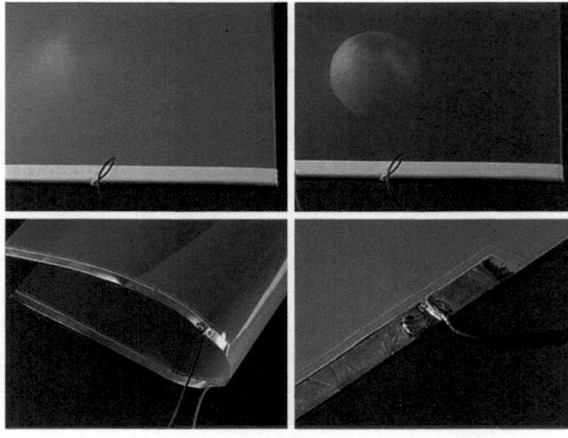

2.4.5 LC Displays

Bistable liquid crystal devices have been extensively investigated because of their unique electrooptic properties, mainly the ability to maintain an image indefinitely without power consumption. The optical requirements of LC devices are: light transmittance almost constant in two different states (on and off); a light transmittance factor of 60–80 % in the active state, and 44–60 % in the inactive state. Other requirements are:

- activation voltage of 30–120 V;
- extremely short response times (10 ms at 200 °C).

Liquid crystals are commonly used on telephone screens, digital cameras, portable computers, and other electronic products. In these screens, the electrical current causes a reorientation of LC molecules parallel to the electric field, changing accordingly the mode in which light passing through the display becomes polarized.

For applications in displays, twisted nematics (TN) are the most commonly used ones thanks to their simplicity (Fig. 2.22).

They are composed of a nematic mesophase placed between two glass plates treated with a polarizing film on the outer surface in order to orient LC molecules parallel to them. The two plates are rotated 90° with respect to each other such that

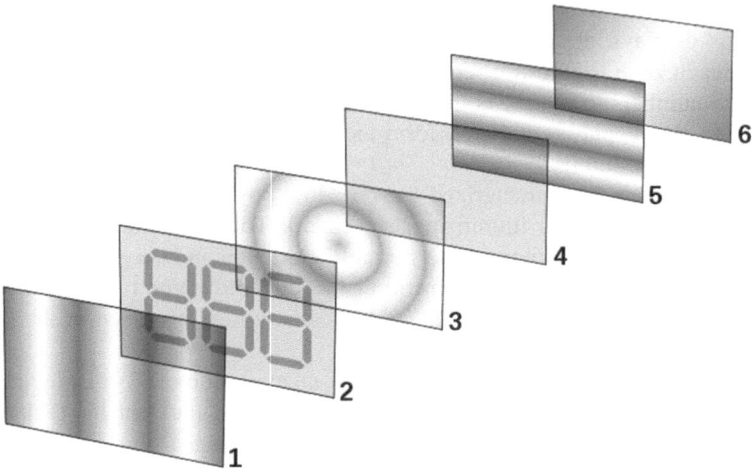

Fig. 2.22 LC display (nematic) with six layers: 1. *Vertical* filter film to polarize the light as it enters; 2. Glass substrate with ITO electrodes. The shapes of these electrodes will determine the *dark* shapes that will appear when the LCD is turned on. *Vertical* ridges are etched on the surface so the liquid crystals are in line with polarized light; 3. *Twisted* nematic liquid crystals; 4. Glass substrate with common electrode film (ITO) with *horizontal* ridges to line up with the *horizontal* filter; 5. *Horizontal* filter film to block/allow through light; 6. Reflective surface to send light back to viewer. Image by ed g2s.talk, through Wikimedia Commons

molecules adjacent to the upper layer are oriented perpendicularly to those adjacent to the lower layer. At the heart of the LC phase, the molecules will therefore tend asume intermediate positions, creating a helicoidal structure similar to that of the cholesteric phase.

In TN displays, the activation of each single point is commanded by an indexing mechanism addressing the row and column, thus each pixel is activated when current passes from both front and back electrodes that are related. Furthermore, in reflecting displays (which make use of the ambient light to become illuminated) there is a mirror placed below the second polarizer. Thanks to this mirror, ambient light is polarized at the first polarizer and enters the liquid crystalline phase. By virtue of the helicoidal form of molecules, the polarization of incident light is rotated 90° so that it can pass also through the second polarizer. It has therefore a subsequent reflection in the mirror, a new rotation of 90° of the plane of polarization induced by the liquid crystalline phase and finally the exit from the first polarizer. The LCD explained as such appears illuminated. If instead an external electric field is applied to the LCD that is perpendicular to the planes of the glass plates, the molecules will tend to orient themselves along the field in spite of the action of anchoring of the two plates themselves. The helicoidal structure is thus broken and the LC phase will not be able to rotate the plane of polarization of the incident light anymore. Since radiation that enters from the top becomes blocked by the second polarizer, the display appears black.

An LCD display is composed of many small zones (or pixels) which appear black or bright depending on whether they are subjected to an external electric field. To this, one adds the effect of coloration previously described. In case of calculator displays or anytime these zones to be lit separately are limited in number, each zone is part of a separate electrical circuit. When, instead, pixels are greater in numbers, a matrix is used where the transparent electrodes located on one side of the display (e.g. back) form the rows and those at the other side (e.g. front) form the columns.

The TN display is characterized by very low power consumption (because it uses the ambient light to illuminate the screen). Thanks to the low cost, it has found numerous applications in the portable computer industry. Among the disadvantages, however, we should mention the quite low contrast and the rather restricted *correct vision* angle (circa 20°) due also to the presence of the layer of transparent electrodes in front of the screen.

The supertwisted nematic (STN) display is an evolution of TN display in which the molecules of the nematic phase (thus the polarized incident light) undergo a rotation of 270° instead of 90°. The principal advantages offered by this technology compared to TN are: much better contrast (about three times) and a larger angle of *correct vision* (about double).

However, they present major problems of birefringence due to the emerging light color can sometimes *shift*. Nevertheless, appropriate methods to correct these drawbacks are available.

The *thin film transistor* (TFT) display is the latest innovation in the field of LCDs. In TFT displays, the addressing of row × column of each single pixel takes place in the display itself and each pixel is activated by an appropriate transistor. Thus, it is no longer necessary to place a series of electrode-rows or electrode-columns while it is sufficient to have a single transparent plate with the function of *grounding element*.

The contrast can thus be quadrupled with respect to STN and the angle of vision becomes slightly wider. It is clear however, that currently the higher complexity of the TFT technology and the higher cost limits their use to prevalently professional applications.

With regard to other LC technologies, cholesteric nematic devices ChLCs (mixed crystals with dichroic molecules), which were developed for a faster reaction, find an advantageous usage in automotive mirrors and in goggles.

Many ways have been developed to produce bistable LC devices over the past years. Among them, a bistable display utilizing an ion-doped smectic-A (SmA) LC is of great interest because of its potential applications as low-power and low-cost display devices, switchable storage devices, and electronic paper.

2.4.6 Liquid Crystal Devices for Windows

In architectural applications, plates with LC devices are prevalently utilized in single plates as filter elements in places such as offices, hotels, restaurants, hospitals, etc. In such places, by maintaining the light transmission unaltered, the windows allow the *visual* division of spaces as frequently needed in meeting rooms, receptions, and environments in which the use requires privacy for a certain period. In the case of external window applications, solutions with double pane windows with air space are preferred. The device is activated by a switch placed in the primary power circuit. The electrodes needed are inserted inside the window frame.

The technology usually applied in windows system of large surfaces is *polymer dispersed liquid crystals* (PDLCs). Among the devices with variable transparency, PDLCs are the ones which are the most commercialized. LCs are applied in commercial switchable glazing or windows and PDLCs, since the 1990s. In 1991 Privalite Inc., for the first time in Europe, introduced a commercial electrooptic glass, a system based on liquid crystals and still available today.

A PDLC, in its simpler shape consist of a laminate system of glass plates which includes a sandwich of two PET (or polyester) substrate layers coated on the inner sides with a transparent conductive oxide, e.g. indium tin oxide (ITO), to serve as electrodes, enclose a polymer matrix in which the LC microscopic molecules dispersed in a polymer matrix are embedded in random order. When electric power is applied, the glass surface appears transparent. When switched off, the surface becomes milky white. In this state, the windows become perfect surfaces for projection and rear projection.

From the simple structure explained above, it is possible to move on to more complex structures with more layers.

For exterior windows, the light transmission is in the range of 60–80 % in the *on* state and 40–60 % in the *off* state. The light transmission can be almost constant in the on/off states. Solar transmission is 79-80 % and 44–60 % in the *on* state. The supply voltage varies from 30 to 100 V and the switching times are extremely short.

Unlike the ECDs, LCDs do not have a capacitive memory, therefore they require a continuous supply of current to remain transparent and a higher voltage for activation.

2.5 Chemochromic Materials

Chemochromic materials are defined as materials which change color in the presence and by the effect of chemical agents. This category of chromogenic materials comprises gaschromic, halochromic, solvatochromic, hygrochromic (Fig. 2.23) and hydrochromic materials (Ritter 2007).

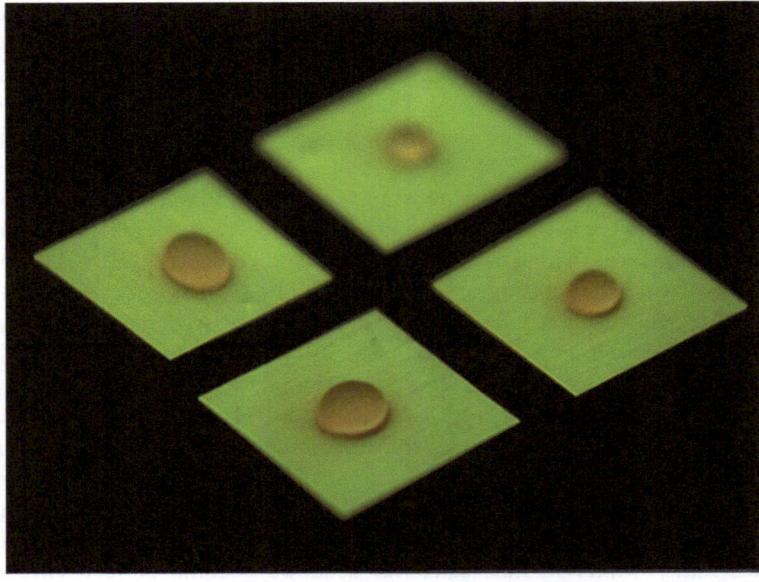

Fig. 2.23 Picture of a polyamide/dye blend, exposed to moisture. *Orange color* (as opposed to *green*) identifies regions of the material that experienced exposure to moisture. Photo of Brent Crenshaw

2.5.1 Gaschromic Materials

Gaschromic materials change color in the presence of gas and they have the capacity to return to the initial state once the cause which led the variation is removed. Until today, many materials have demonstrated gaschromic behavior. Among them, notable ones include tin oxide (SnO_2) which reacts to flammable gases (Kamimori et al. 1994); palladium, which is sensible to hydrogen (H_2) (Hofheins et al. 1995); titanium tin oxide (Ti, Sn)O_2 which react with carbon monoxide (CO), propane (C_3H_8), ethanol (C_2H_5OH) and H_2 (Arakawa et al. 1999); tungsten oxide (WO_3) which is the most studied gaschromic material and reacts with hydrogen disulfide (H_2S) (Dwyer 1991) and with H_2 at ambient temperature, turning blue as observed by Khoobar as early as 1964 (Patil 2000).

Through the application of thin layers of these materials (nanocoatings) by chemical vapor deposition, optical coatings have been developed for applications in window systems with the function of reducing thermal energy loss or solar overheating of buildings, reducing heating or cooling costs. Chemochromic windows or other transparent building components, when exposed to gas, become activated thanks to the presence of a catalyst layer. Unlike electrochromic devices, gaschromic devices do not need electrode layers. Gaschromic windows transmit more than 80 % of solar radiation (Patil 2000).

In 2001, collaboration between Fraunhofer Institute for Solar Energy Systems ISE and the window manufacturer Interpane (Lauenförde, Germany) resulted in the realization of a pilot production plant for insulating and dynamic windows and the assembly of insulating windows which darken with the push of a button (Fig. 2.24). The gaschromic window system is suitable to efficiently substitute the traditional concealed roller blind or Venetian blind systems since it is possible to

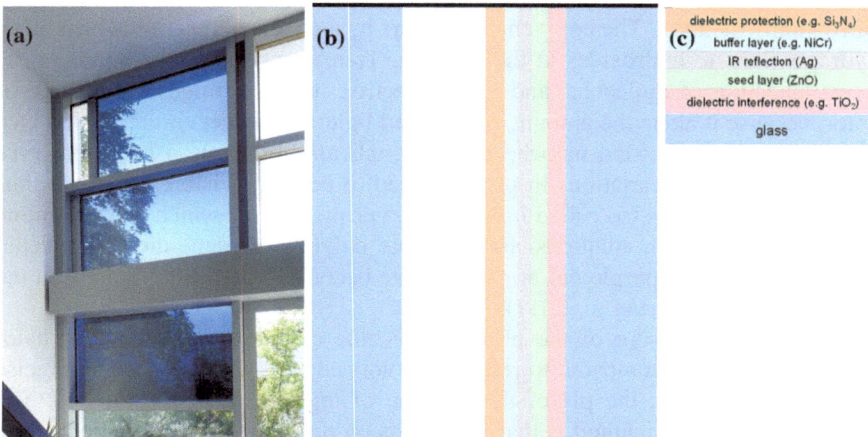

Fig. 2.24 Darkening effect of gaschromic windows and various layers deposited on glass needed for the chemochromic technology. *Courtesy* Fraunhofer ISE

reduce the transmittance by 75 %. Gaschromic windows involve a double pane construction with a gap between the two panes. One of the surfaces is coated with tungsten oxide and a thin layer of catalyst on top. The thin layers, invisible to the naked eye, transform regular glass into a gaschromic one. When a small quantity of hydrogen is pumped into the gap between the panes, tungsten oxide obtains a blue tinge but remains transparent. When oxygen is introduced, the reaction is reversed and the glass becomes clear again. A small electrolysis unit is hidden in the facade which produces the hydrogen gas and a miniature pump diffuses the desired gas mixture into the cavity. The hydrogen concentration is very low and does not pose any risk. The visual perception during the switching of color is fascinating: just like in a solar eclipse, the level of ambient light changes in a uniform fashion in a few minutes, in total absence of noise (Lampert 2004; Georg 2002).

Science fiction novels and movies are a great inspiration for innovation. In one of the episodes of the sci-fi TV series (*Battlestar Galactica* 2004), the heroine *Starbuck* finds herself stranded on a barren moon. She needs to understand whether the atmosphere contains oxygen because her suit oxygen runs low. She takes out a pen-shaped device and slides it open. The inner part of the pen which is red remains red, there is no oxygen in the atmosphere; she becomes disappointed. Then she finds the *Cylon Raider* that attacked her. She manages to get into the raider and tries her oxygen sensor once again. Luckily, there is an oxygen supply in the raider and the red indicator inside the pen turns green. The director may not know it but this is an oxygen sensor using a *gaschromic* material!

2.5.2 Halochromic Materials

Halochromic materials change color according to the changes in the acidity of the surrounding medium. The color change occurs by a chemical reaction which binds hydrogen ions to hydroxides in the solution. Transformation of the bonds in the interconnection of molecules and in the electron flux determines the change of color, because it alters the amount of absorbed light and hence, emitted color. Due to their special properties, halochromic materials are frequently used in places or situations in which variations in pH is needed to be made visible. They are thus used as pH indicators (so called *litmus paper*) or pH sensors which can determine changes of acidity in solutions. For example poly(3,4-ethylene-dioxythiophene) (PEDOT)) which is purple red at neutral state becomes blue or transparent in the oxidized state (Tiwari et al. 2011).

To obtain a measure of the pH variation, the color obtained with the halo-chromic material is compared to reference colors each of which indicate a precise pH value. This way, the pH of the solution of interest may be estimated. The disadvantage of this method is that it depends on the subjective evaluation of the user and the sensibility of the human eye to colors.

Currently, many researchers are engaged in developing applications using halochromic dyes in textile products with the aim of realizing flexible pH sensors, made with different techniques and for use in various fields.

Several studies demonstrated that halochromic dyes can be applied on conventional fabric with normal techniques and that the response of these sensors depends on the density of the fabric. The technique of coating through sol-gel appears to be one of the best among those attempted because the applications show a clearly visible color change upon variation in environmental acidity and short response times. The coloration of halochromic nonwoven fabric made of nanofibers can instead be obtained directly during the formation of the fibers. The response of these sensors is rapid thanks to the high porosity of the substrate made of nanofibers. However, it should be noted that the halochromic behavior of colorants applied on a textile matrix is different from their behavior in solution because an interaction occurs between fibers (Van der Schueren and De Clerck 2013).

Halochromic textiles pose unique opportunities for simple and smart solutions. For example, in the medical sector, halochromic bandage can be used for the monitoring of patients with burns. Since pH changes during the healing process of the skin, with textile bandages which employ halochromic molecules, it is possible to monitor the level of healing without having to remove the bandage and cause pain or discomfort to the patient (Osti 2008 in Zong 2013). Fabrics treated with halochromic dyes have also been used in geotextiles or protective clothing that measures pH alteration in air in real time. Another use of halochromic colorants is the production of paints for metals which change color as a result of corrosion of the metal underneath as in the case of the formation of rust in steel structures. In the building sector, this application could have a tremendous market.

A particular type of halochromic materials are the oxazines. These molecules are being studied to obtain *paint with adjustable color* that change color after application so that their color corresponds to the color of adjacent areas or for decorative purposes. Colorants obtained with these molecules are characterized by a high controllability of color. They change color upon the application of chemical agents and may maintain a specific color for a prolonged period of time such as several months. The color persists indefinitely unless a solution of basic pH is applied. The structure of the compounds can be regulated in order to optimize their reversibility (Tomasuolo and Raymo 2010).

2.5.3 Hygro-Hydrochromic Materials

Hygro-hydrochromic materials change color in response to the presence of moisture or to contact with water. They form a category of a larger group: materials which present the phenomenon called *solvatochromism*, typical of some chemical substances which are sensitive to a given solvent (liquid or gas).

Solvatochromism depends on the particular interaction between the molecules of the substance and the molecules of the solvent. The substance which contains chromophore groups is sensitive to the polarity of the solvent, which functions like a constant electrical field and determines the effects on the spectroscopic properties of the substance, hence a change in color.

In the field of chemical research, solvatochromism is used for environmental sensors, in the analysis of probes with the capacity of determining the presence and the concentration of a solvent, and in molecular electronics for the construction of molecular switches.

In industrial products, hydrochromic materials are used for special inks in the form of pigments, which in many cases, act similar to thermochromic ones: they are normally white and opaque but become transparent in the presence of water.

In most of the applications which were realized until today, these materials were applied on surfaces which already have a colored layer printed on them, forming a fine white film which rejects light waves, impeding them from reaching the printed image. At the moment in which the surface is wetted with water, the film which used to diffract the light acquires a viscosity such that it becomes permeable to electromagnetic frequencies and lets the light waves to be filtered through, making visible the color image lying underneath. When the surface is dried, the ink film returns to its light impermeable condition, becoming white and opaque again.

The best results with hydrochromic inks in terms of applicability and durability are obtained when the ink is applied on dry and smooth surfaces. Various types of substrates can be used: soft sheet vinyl, paper, coated paper, styrene sheet, polyester fabric, soft PVC, etc. The inks can be applied by screen printing (contact print) or spray coating, followed by passing it through a forced hot air tunnel, as in the production of printed shirts.

Hydrochromic materials have been used in the textile sector to achieve dynamic patterns on textiles. Later, they were used in order to realize clothing and bathing suits such as those designed by Amy Winters (Sect. 5.1), umbrellas such as Squidarellas by Viviane Jaeger and Emma-Jayne Parkes (Fig. 2.1), and tableclothes like Underfull by Kristine Bjaadal (Fig. 2.25), a prototype developed in collaboration with Smart Textilestextile[6] in 2005. At first sight, Underfull appears like a traditional paisley patterned tablecloth; monochromatic, decorated with floral designs with a shiny-opaque effect. However, right when a cup of water accidentally tips over and the tablecloth becomes wet, the hidden graphic design reveals itself by becoming colored. Thus, an unpleasant incident at the table transforms into a positive experience as in the popular Italian saying "accidentally spilling wine on the table brings good luck".

---

[6] *Smart Textiles* is the multidisciplinary research center for innovation, development, design, and production of the next generation textiles. Born in Sweden, it supports the knowledge of cluster of textile firms in the Borås region. In the laboratory environment, the center is supported by the University of Borås (the School of Textiles and the School of Engineering), Swerea IVF/Textiles and Plastics, and SP Technical Research Institute of Sweden. http://www.smarttextiles.se.

Fig. 2.25 Kristine Bjaadal, Underfull tableclothes, concept developed during her MA study at Oslo National Academy of the Arts, 2009. *Courtesy* Kristine Bjaadal

2.6 Mechanochromic Materials

Mechanochromic materials display a change in color or transparency with a change of mechanical stress applied to them. These smart materials can respond to compressive, tensile, or more complex forms of stress. The term *piezochromic* is also used in the literature to describe the same group of materials.

Compared to other types of chromogenic materials such as thermochromic or photochromic materials, mechanochromic materials have not received much attention so far, partly because the related technology is not mature yet. Even patents related to mechanochromic materials are relatively small in number (Ferrara and Bengisu 2013).

Mechanochromic phenomena have been observed in various polymers and inorganic materials. Polymers such as poly(di-*n*-alkylsilanes), poly(3-alkylthiophene) (Chen et al. 2012), poly(3-dodecylthiophene) (Seeboth et al. 2011), and spiropyran (Potisek et al. 2007) were demonstrated to exhibit mechanochromic activity. PMA-spiropyran solutions (PMA-SP-PMA) were subjected to ultrasound at 6–9 °C which caused the colorless solution to turn pink due to the mechanochemical ring opening of the spiropyran molecules. Exposure to ambient light and temperature for 40 min caused the color to disappear. Recently Seeboth et al. (2011) developed a polymeric mixture that involves cholesteryl derivatives. This material was able to respond to very low levels of pressure (several bars) and switch reversibly from red to green color. The ability to obtain a reversible mechanochromic effect at several bars is an important achievement for practical and economically feasible applications.

Chen et al. (2012) used computational analysis to develop new mechanochromic compounds using anthraquinone imide (AQI) derivatives with electron donating substituents. Three mechanochromic polymers with different color transformations were developed, i.e. orange to deep red, dark purple to black, and green-yellow to red. A pressure of 900 MPa was applied to obtain the mechanochromic transformation. The color change was irreversible in these polymers. Mechanochromic luminescence was also observed in the synthesized compounds.

Some inorganic materials also exhibit mechanochromism, including LiF and NaCl single crystals, $CuMoO_4$, and palladium complexes. The Ni (II), Pd (II), and Pt(II) dimethylglyoxime complexes change color under pressures of 6,300–15,000 MPa (Takagi 2004). SmS undergoes a semiconductor to metal phase transformation under a pressure of 0.65 MPa, along with a mechanochromic response (Jin and Tanemura 1999). The α to γ phase transition of $CuMoO_4$ requires a pressure of 250 MPa, reversibly modifying its color from green to brownish red (Hernandez 1999). The pressure needed to induce the phase transition in $CuMoO_4$ was decreased significantly by substituting a small portion of Mo by W. A green to red transition occurs in the compound $CuMo_{0.9}W_{0.1}O_4$ at a pressure of 20 MPa. In other words, a simple push with a finger can cause a phase transition and the accompanying color change in the new material (Gaudon et al. 2007).

Products employing mechanochromic materials are currently at the research and development stage in most of the cases. A climbing rope was developed with the help of a mechanochromic dye at the Smart Structures Research Institute, Strathclyde University, which was able to detect critical levels of strain and signal the user, through a color change, that the rope is damaged and should be discarded (Goddard et al. 1997). Another interesting project is on flexible materials with stress-sensitive color. Researchers from the University of Cambridge and Fraunhofer Institute have developed what they call a "polymer opal". While natural opals display a multitude of colors due to silica spheres that settle in different layers, polymer opals consist of one preferred layer of nanoparticles, resulting in uniform color (University of Cambridge 2013). A remarkable feature of polymer opal is the intense colors which are based on the same physical principles found in nature, similar to that found in some beetles, butterflies, and peacocks. This type of color is known as *structural color* because it results from the diffraction of light due to the structural layout of nanoscale layers. An important advantage of structural color is that it is not dependent on time and temperature as long as the structure is preserved. This means that materials which display structural color would not fade over time, as do most painted or printed advertisements. Polymer opals can be stretched and twisted because the base material has a rubbery consistency. When stretched, the spacing between the nanospheres increases, resulting in a color change to the blue range of the spectrum (Fig. 2.26). Compressing the material results in a color shift towards red, and when released, the material returns to its original color.

The patent literature describes various conceptual applications for mechanochromic materials. Many of these applications are related to stress sensing and failure detection. One of the major driving forces for research in this field is the

Fig. 2.26 Polymer opal gradually turns to *blue* when stretched and returns to its original color, in this case *green. Courtesy* University of Cambridge

aerospace industry due to the high maintenance and inspection cost of air and space vehicles. Several patents propose applications where mechanochromic materials signal an unusual level of stress, which may be due to corrosion, overloading, or crack formation. An indirect approach involves microcapsules or hollow fibers which contain a colored substance. Under high stress, the hollow fibers or microcapsules break up, releasing the colored ingredient in order to signal a critical condition of the related component. One difficulty with this approach is to obtain a homogeneous distribution of these fibers or microcapsules, which is necessary for it to function successfully. Two part systems have also been developed where a dye and an activator, contained in separate microshells, mix and react in order to result in a colored substance when overload, cracking, or damage occurs. A direct approach is to use mechanochromic polymers such as diacetylene segmented copolymer or spiropyran for the same application (Potisek 2010), which is much simpler to manufacture and apply to the desired system.

A smart toothbrush was invented by Unilever. The body of the toothbrush uses a mechanochromic liquid crystal cholesterol ester which indicates to the user whether correct or excessive force is being applied during brushing. The product aims to prevent damage caused to teeth and gums due to brushing too hard (Davies et al. 2001; Savill 2002).

Another application of mechanochromic materials was described for industrial roll covers (Lutz 2007). Such rolls are employed in papermaking, printing, and coating industries. Typically electronic sensors are used to prevent uneven or extreme pressures on these rolls, which are damaging them. Mechanochromic materials are proposed as a more feasible alternative to electronic sensors since electronic sensors typically require a power source, some communication equipment, data processing, and maintenance. Mechanochromic coatings would not require any of these. However, the use of mechanochromic coatings requires visibility: the whole roll surfaces should be visible during operation, which may not be possible for all such systems.

In addition to possible applications discussed so far, we propose some applications which might expand the horizon of chromogenic materials (Ferrara and Bengisu 2013).

Mechanochromic Paint for Tennis, Volleyball, Basketball, and Similar Courts
Those who play or watch tennis matches would appreciate how difficult it is to rule whether a ball served at 180 km/h that hits close to the service line or side line is in or out. This problem was initially addressed by line judges and electronic sensors. An electronic line judge was first used in a tennis tournament in 1977 (New Scientist 1977). Because of the high cost and occasional breakdowns, other options have been also tried. Since 2006, USTA uses instant replay technology along with a player challenge system for grand slam tournaments (USTA 2013). Players can challenge a line call and an official replay is used to see where the ball dropped exactly. Although this technology is more dependable than electronic line judges, it is costly and interruptions occur during the game. Our solution to this problem is relatively simple. The court is painted with a paint which contains mechano-chromic pigments with the ability to sense the force applied by the ball and reversibly change color from its normal color (green or blue) to a contrasting color such as red or orange. The mark left by the ball should fade away slow enough for the referee and players to decide whether it was in or out, for example 60s. It is sufficient to use the mechanochromic paint for the lines and near them, since the indecision only occurs in these regions and also because footprints due to the mechanochromic effect all over the court would cause confusion. The same type of paint could be used for other courts such as squash, volleyball, and basketball courts (Fig. 2.27).

Mechanochromic Bathroom and Kitchen Scales
There are many types of scales in the market today, used for different weighing purposes. Two types of scales used at homes are bathroom and kitchen scales. A common and elegant design used for bathroom scales involves a glass platform and a digital display. Unlike this electronic scale, a mechanochromic scale would not need any batteries, mechanical or electronic parts. The product concept consists of a glass plate, under which there is a graphic design which involves me-chanochromic ink. This image will show the corresponding weight of the person with the help of colors, graphic, or numbers. The mechanochromic image can be

Fig. 2.27 *Tennis ball* leaving a *red mark* on chromogenic paint

printed on a polymeric backing plate or it can be applied as a separate film. A similar approach can be used for a kitchen scale. However, in this case, a container is needed to contain the sugar, flour, or milk. Thus, a special container with a small base may be designed to provide more visibility for the mechanochromic graphic that must be seen by the user.

Mechanochromic Packaging

The use of a mechanochromic label or surface indicating overload conditions could be a useful application under different circumstances. Overloading of packaged items could be a problem especially with fragile, liquid, or gas contents. Some common situations where overload might occur are stacking of many boxes on top of each other in warehouses or supermarkets, which might cause damage to the lowermost box, or overloading of an individual box which, when lifted, may cause breakage or tearing of a box and which may hurt the person who lifts it. Mechanochromic labels or surfaces can be used to indicate an overload condition in the box or package. A related application would be tamper-evident closures. A mechanochromic film can be used in the closure area of a food or medicine container. Tampering or accidental damage to the closure area would be made visible through the color change of the film. A 'similar idea was mentioned specifically for pharmaceutical bottles and jars (Kirk–Othmer 2013).

Certain conditions in canned foods result in slight or obvious swelling or bulging of the container. For example, enzymatic action, non enzymatic browning, and microbial growth cause carbon dioxide evolution and spoilage of the ingredients (Downs and Ito 2001). A mechanochromic coating on the can could be designed and developed to warn the consumer about the possible spoilage of the canned ingredients.

Mechanochromic materials can also indicate a decrease of pressure if the packaging and the mechanochromic substance is designed accordingly. This effect could be used in pressurized tennis ball cans or other pressurized packaging. High quality tennis balls are kept under ~ 2 atm pressure so that the balls maintain their pressure while they are stored. A visible label on the packaging could be used to detect a fall in pressure which might be caused due to a defective package or because it was opened, to warn the customer about the condition of the product.

Mechanochromic Game

This is a mystery game intended for children, age 8–13. The game involves several cube blocks made of an elastomeric material such as synthetic rubber (polyisoprene or silicone rubber). Alternatively, wooden or plastic cubes can be covered with an elastomeric material. Each face of each block is printed with a mechanochromic image. When players rub two blocks to each other, a symbol appears temporarily on the faces which are being rubbed to each other, because of the pressure. One set of the blocks is the key to the symbols. The players try to find out what types of symbols are written on the blocks and what they mean. When they put together all symbols on all sides of one or more blocks and when they find out which letter or number they correspond to, they solve the mystery. This game could be played by one or two players or by two or more teams.

2.7 Biochromic Materials

The use of chromogenic media for the detection of pathogens is not new but rather one of the commonly used methods for this purpose today. Chromogenic culture media, developed since 1990, target microbial enzymes and release colored dyes upon hydrolysis. This approach results in the formation of colored colonies which stand out against the background flora and allow the identification of pathogens easily (Perry and Freydière 2007). However it requires the growth of a single cell into a colony and the detection procedures may last up to several days (Ivnitski et al. 1999; Velusamy et al. 2010). In response to disadvantages of conventional methods employed for pathogen detection, effort is directed to the development of biosensors with rapid detection capability of even small numbers of pathogens. *Biochromism* is a term used to define a color change due to a biochemical reaction. Biochromic materials have been studied for their biosensor capabilities in the form of simple structures such as membranes, liposomes, and Langmuir–Blodgett (LB) monolayers. Biological membranes are critical to biological processes such as transport, signal transition, and molecular recognition (Song et al. 2002). Cell membrane and related structures are mimicked in order to facilitate the design and development of biosensors. Conjugated polymers such as polydiacetylene (PDA) and polythiophene (PT) are commonly investigated for these purposes (Bamfield and Hutchings 2010). PDA has the ability to self assemble into organized vesicles and films and to show a drastic color change from blue to red under heat, mechanical stress, or upon molecular recognition of certain pathogenic agents (Scindia et al. 2007; Song et al. 2002; Su et al. 2005). PT is an organic polymer and shows chromic transitions upon excitation by heat, metals, chemicals, and proteins. PT-based materials functionalized with carbohydrates such as sialic acid and mannose have been produced to detect bacterial toxins. A red shift in UV absorbance was registered upon detection of toxins (Ahmed and Narain 2011). A recent study investigated the direct application of surfactant functionalized PDA vesicles to detect common bacteria *Staphylococcus aureus* and *Escherichia coli* in a culture medium and in apple juice. Bacterial detection of PDA vesicles coated on cellulose strips was also analyzed. Interestingly, apple juice alone could trigger a color shift in the films but the color shift was stronger in the presence of bacteria (dos Santos Pires et al. 2011).

The colorimetric response of PDA films was shown to occur within several minutes upon contact with the solution containing selected bacteria. The color change is believed to occur due to (a) the interaction of bacterially secreted membrane-active compounds such as bacteriocins and the receptors and/or (b) insertion of hydrophobic peptides and proteins into the PDA membrane (Scindia et al. 2007; Su et al. 2005). Cell surface receptors incorporated onto PDA vesicles or LB monolayers interact with specific viruses or toxins and a dramatic color change occurs from blue to red. The color change is *irreversible* in PDA supramolecular assemblies. The mechanism involved in the color change is not clear, although it is believed to be due to distortion of the backbone of the PDA polymer (Song et al. 2002). The color intensity typically depends on the species, strains, growth rate and population of pathogens. Biosensor signal amplification can be achieved through increasing the solution pH or phospholipid content in the mixed lipid vesicles (Arshak et al. 2009).

Although biochromic materials are relatively new, some applications already emerge in the industry. One of the target markets of biochromic technology is the food industry, specifically in the detection of foodborne bacteria. The development of sensors suitable to be used on packaging of any food product that may contain bacteria or toxins is an important target for health and quality concerns. There are many technologies being developed for such a purpose but biochromic devices have a high potential to offer important advantages such as easy adaptability to packaging and no external energy requirement. One such application is described in detail in Chap. 5. Researchers at Berkeley National Laboratory developed biochromic sensors (also called colorimetric biosensors) for detection of viruses, bacteria, parasites, neurotoxins, and other pathogens (Berkeley Lab 2009; Charych et al. 1993, 1996). The engineered membranes consist of chemically or biologically specific ligands attached to a PDA backbone. These biochromic sensors have potential applications such as rapid test kits for influenza and other diseases, detection of foodborne bacteria such as *E. coli*, drug research and development, detection of DNA hybridization, and detection of pollutants.

Fig. 2.28 Chameleons are famous for their striking colors and their ability to change their skin color. *Courtesy* Stefanie Trautweiler

2.8 Dynamic Color in Nature

Dynamic color is not a phenomenon unique to synthetic chromogenic materials alone. Nature hosts many examples of animals and plants which change color for different reasons. Leaves of deciduous trees change color from green to yellow and red during fall. This change is caused by the unmasking of carotenoid and anthocyanin pigments when chlorophyll production stops (Lee 2007). However, this change is very slow and not comparable to the speed of color change in chromogenic materials. On the other hand, there are many animals which display a rapid change in their skin color. For example, chameleons are famous for their ability to change color in a short time and display a variety of beautiful colors and patterns (Fig. 2.28). Typically they change color to communicate with other chameleons especially during courtship (West 2009).

Octopuses, squids, and flatfish such as sole and flounder adapt their color to their surroundings to camouflage themselves effectively (Armstrong 2001). The color change occurs in the skin due to pigment containing cells called *chromatophores* in minutes or seconds. There are two basic mechanisms which lead to color change in the skin. In cephalopods (octopuses, squids, cuttlefish), radial muscles are relaxed or contracted by active neurophsiological control from the brain. Relaxation of the muscles shrinks the pigment sacs of chromatophores and concentrates the pigments. When the muscles contract, pigment sacs open up and the pigments are dispersed. The second mechanism involves simple dispersion of pigment within the chromatophore, which occurs in amphibians such as frogs. The movement of pigment-containing organelles is under neural or endocrine control (Kay 1998; Stevens and Merilaita 2011).

An additional feature seen in cephalopods is the creation of patterns for communication and camouflage. Various patterns can be created on the skin of these animals but the number is fixed, not unlimited. These patterns are created not only by chromatophores (containing mostly brown, red, and yellow pigments) but also by leucophores (white scattering color cells), iridophores (directional structural reflectors forming red, green, and blue color), and skin muscle. For example, the cuttlefish *S. officinalis* displays three basic patterns which were categorized as *uniform, mottle*, and *disruptive*. Out of the 34 chromatic components it possesses, 11 are used to construct variations of the disruptive pattern (Stevens and Merilaita 2011).

Although a great deal of research has been conducted so far on the adaptive camouflage and dynamic color skills of animals, nature is full of opportunities for further investigation and learning something new. New discoveries and answers lead to new questions. Meanwhile, tools of biomimetics would be quite useful for designers and engineers in order to develop ideas, materials and devices for new chromogenic technologies.

As has been demonstrated recently, apart from natural pigments and the chemical response of materials, the chromogenic effect may also be originated by the physical structure of surfaces. The study on nanoscale photonic structures of certain living organisms, such as the tropical beetle *Dynastes hercules* and *Hoplia coerulea* demonstrated that the optical properties of some surfaces, such as the shell of the insect, derive from the geometrical structure, smaller in scale than a micrometer, which interferes with light.

The male beetle of *Dynastes hercules* is usually greenish in color, but when the humidity of air exceeds 80 %, it transforms to black. This phenomenon was termed *hygrochromism* by the researchers, which fits the general meaning of the term well. However it should be noted that the mechanism of color change is completely different from any chromogenic phenomenon observed with natural or synthetic organic molecules or inorganic colorants discussed throughout this book. The water penetrates the three dimensional structure of the cuticle surface of the beetle and changes the optical properties. Regardless of why this happens, whether it is better to blend in during the day or at night with the dominant colors or for the thermoregulation during the day, the phenomenon is determined by the three-dimensional morphology. In more detail, the greenish color, visible during the dry state, derives from a three dimensional layer with open pores located ~ 4 µm below the cuticle surface. The colored structure is made of a network of horizontal filaments, arranged in thin layers parallel to the surface of the cuticle, stiffened perpendicularly by a structure of vertical cylindrical elements (similar to pillars). The diffraction of broadband light in this layer generates greenish color. The backscatter disappears when this structure is infiltrated by water, reducing the difference in the refractive index (Rassart et al. 2008).

Nature is full of animals which display surprising vivid, brilliant colors. Color in animals is mostly associated with pigments. However, in certain animals, color is caused without the help of pigments. This second type of color is caused by the structural arrangement of very fine, typically nanoscale matter and interaction of this special structure with light. This phenomenon is called *structural color*. This is

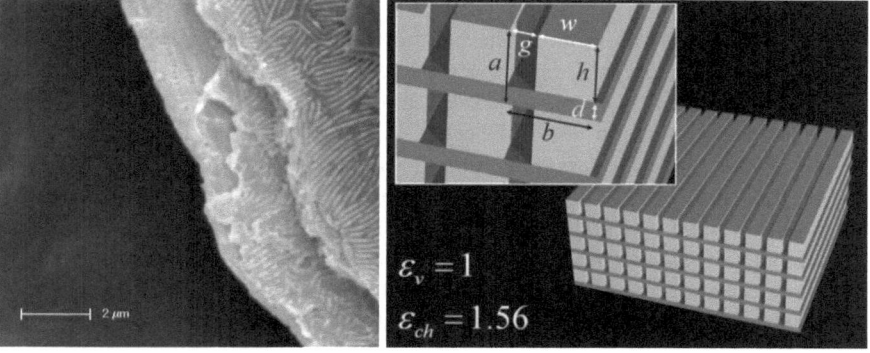

Fig. 2.29 *Left* SEM images analysis the structure of the beetle *Hoplia coerulea*. *Right* Idealized coloring structure of the beetle. The distances *b*, *g*, *w*, *a*, *h*, *d* are deduced from SEM analysis and can be used to evaluate the dominant reflected wavelength. Research of Vigneron et al. (2005) at Research Center in Physics of Matter and Radiation, University of Namur, Belgium

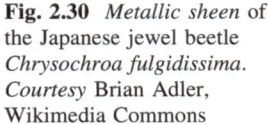

Fig. 2.30 *Metallic sheen* of the Japanese jewel beetle *Chrysochroa fulgidissima.* *Courtesy* Brian Adler, Wikimedia Commons

observed for example in peacock feathers, some butterfly wings, and beetles. An important feature of structural color is its angle dependency, also called *irides-cence*. The color changes when the viewing angle of the observing eye is changed. In that sense, this is a dynamic effect but not a chromogenic phenomenon. Structural colors in nature are believed to occur due to five fundamental optical processes. These are thin film interference, multilayer interference, diffraction grating, photonic crystals, and light scattering. A combination of some of these five effects and pigmentation is also possible. For example, the blue iridescent color of *Morpho* butterflies is explained through the combined action of inter-ference, diffraction of light, and pigmentation (Kinoshita and Yoshioka 2005).

In the case of the male *Hoplia coerulea,* noted for its iridescent blue-violet coloration, color has its roots in the photonic structure of the dorsal part of the insect's body (Fig. 2.29). This is made of very fine, superposed layers of chitin which are supported by filaments parallel to each other (Vigneron et al. 2005).

The metallic reflection and iridescence in beetles such as the Japanese jewel beetle (*Chrysochroa fulgidissima*) (Fig. 2.30), flower beetle (*Protaetia pryeri*), frog-legged leaf beetle (*Sagra femorata*), gold scarab beetle (*Chrysina resplen-dens), and Plateumaris sericea* (a leaf beetle) is explained in terms of multilayer interference (Kinoshita and Yoshioka 2005; Lenau and Barfoed 2008). Research is dedicated to reproduce materials and architectures that provide similar iridescence effects with the aim of applying them in industrial products. Details on the topic of structural color can be found in two excellent reviews (Kinoshita and Yoshioka 2005; Lenau and Barfoed 2008).

References

Ahmed M, Narain R (2011) Glycopolymer bioconjugates. In: Narain R (ed) Engineered carbohydrate-based materials for biomedical applications. Wiley, Hoboken

Arakawa S, Mogi K, Kikuta K, Yogo T, Hirano S (1999) J Am Ceram Soc 82:225

Armstrong M (ed) (2001) Aquatic life of the world. Marshall Cavendish Corp, New York

Arshak K et al (2009) Conducting polymers and their applications to biosensors: emphasizing on foodborne pathogen detection. IEEE Sens J 9:1942–1951

Bamfied P, Hutchings MG (2010) Chromic phenomena: technological applications of colour chemistry. R Soc Chem, Cambridge

Battlestar Galactica (2004) Season 1, episode 5, You can't go home again. https://en.wikipedia.org/wiki/You_Can%27t_Go_Home_Again_(Battlestar_Galactica). Accessed 10 July 2013

Berkeley Lab (2009) Colorimetric and fluorescent sensors for rapid and direct detection of influenza, E. coli and other analytes. http://www.lbl.gov/tt/techs/lbnl0965.html. Accessed 8 July 2013

Billah SM et al (2008) Direct coloration of textiles with photochromic dyes, part 2: the effect of solvents on the colour change of photochromic textiles. Color Technol 124:229–233

Berzowska J (2005). Electronic textiles: wearable computers, reactive fashion, and soft computation, textile, vol 3(1). Berg, UK, pp 2–19

Canal C et al (2008) Atom-sensitive textiles as visual indicators for plasma postdischarges. Appl Surf Sci 254:5959–5966

Charych DH, Nagy JO, Spevak W, Bednarski MD (1993) Direct colorimetric detection of a receptor-ligand interaction by a polymerized bilayer assembly. Science 261:585–588

Charych et al (1996) A 'Litmus test' for molecular recognition using artificial membranes. Chem Biol 3:113–120

Chen F, Zhang J, Wan X (2012) Design and synthesis of piezochromic materials based on push–pull chromophores: a mechanistic perspective. Chem Eur J 18:4558–4567

Corredor CC et al (2006) Two-photon 3D optical data storage via fluorescence modulation of an efficient fluorene dye by a photochromic diarylethene. Adv Mater 18:2910–2914

Davies RH, Savill DG, Jones P (2001) Toothbrush. US Patent 6,330,730, 18 Dec 2001

Demus D, Richter L (1979) Textures of liquid crystals. Verlag Chemie, Weinheim

Dos Santos Pires AC et al (2011) A colorimetric biosensor for the detection of foodborne bacteria. Sens Actuators B 153:17–23

Downs FP, Ito K (eds) (2001) Microbiological examination of foods. Am Public Health Assoc, Washington DC, p 584

Durasevic V, Parac Osterman D, Sutlovic A (2011) From murex purpura to sensory photochromic textiles. In: Hauser P (ed) Textile dyeing. InTech, Croatia. http://www.intechopen.com/books/textile-dyeing/from-murex-purpura-to-sensory-photochromic-textiles. Accessed 25 June 2013

Dürr H, Bouas-Laurent H (eds) (1990) Photochromism: molecules and systems. Elsevier, Amsterdam

Dwyer DJ (1991) Sens Actuators B5:155

Ferrara M, Bengisu M (2013) Intelligent design with chromogenic materials. J Int Colour Assoc (to be published)

Fraunhofer ISE (2013) Photoelectrochromic windows. http://www.ise.fraunhofer.de/en/areas-of-business-and-market-areas/energy-efficient-buildings/facades-and-windows/glazing/photoelectrochromic-windows. Accessed 27 July 2013

Fukuda Y (ed) (2007) Inorganic chromotropism: basic concepts and applications of colored materials. Springer, New York

Gaudon M, Deniard P, Demourgues A et al (2007) Unprecedented "one-finger-push"-induced phase transition with a drastic color change in an inorganic material. Adv Mater 19:3517–3519

Georg A (2002) A chemical solar eclipse in your window. Press release, Fraunhofer Institute of Technology.http://www.archiv.fraunhofer.de/archiv/presseinfos/pflege.zv.fhg.de/english/press/pi/pi2002/04_pi_fenster_br.html. Accessed 22 July 2013

Gillaspie DT, Tenent RC, Dillon AC (2010) Metal-oxide films for electrochromic applications: present technology and future directions. J Mater Chem 20:9585–9592

Goddard NDR, Kemp RMJ, Lane R (1997) An overview of smart technology. Packag Technol Sci 10:129–143

Gordon W (2004) Polymers and liquid crystals (virtual textbook). http://plc.cwru.edu/tutorial/enhanced/files/textbook.htm. Accessed 17 Dec 2012

Granqvist CG (1995) Handbook of inorganic electrochromic materials. Elsevier, Amsterdam

Granqvist CG, Green S, Niklasson GA, Mlyuka NR, von Kræmer S, Georén P (2010) Advances in chromogenic materials and devices. Thin Solid Films 518:3046–3053

Gray GW, Goodby JW (1986) Smectic liquid crystals textures and Structures. Leonard Hill, Glasgow

Gu Y, Cao J, Wu J, Chen L-Q (2010) Thermodynamics of strained vanadium dioxide single crystals. J App Phys 108:083517

Hallcrest (2013) Thermochromic Ink used for Coors cold activated bottle label. http://www.hallcrest.com/coors.cfm. Accessed 1 July 2013

Hampp (2000) Bacteriorhodopsin: mutating a biomaterial into an optoelectronic material. Appl Microbiol Biotechnol 53:633–639

Hernandez D, Rodriguez F, Garcia-Jaca J et al (1999) Pressure-dependence on the absorption spectrum of $CuMoO_4$: study of the green to brownish-red piezochromic phase transition at 2.5 kbar. Phys B 265:181–185

Hofheins B, Lauf R, Felton J (1995) Sensors 12:131

Hu J (2010) Adaptive and functional polymers, textiles, and their applications. Imperial College Press, London

Hu W et al (2010) Magnetite nanoparticles/chiral nematic liquid crystal composites with magnetically addressable and magnetically erasable characteristics. Liq Cryst 37:563–569

Ivnitski D, Abdel-Hamid I, Atanasov P, Wilkins E (1999) Biosensors for detection of pathogenic bacteria. Biosens Bioelectron 14:599–624

Jin P, Tanemura S (1999) Manufacturing methods of samarium sulfide thin films. US Patent 6,132,568, 17 Oct 2000

Jones J (2007) Fashion trends: how popular style is shaped. Capstone Press, Minnesota

Kay I (1998) Introduction to animal physiology. BIOS Scientific Publishers, Oxford

Kamimori M, Suzuki K, Ohya Y, Takahashi Y (1994) Jpn J Appl Phys 33:6680

Kanu SS, Binions R (2010) Thin films for solar control applications. Proc R Soc A 466:19–44

Kinoshita S, Yoshioka S (2005) Structural colors in nature: the role of regularity and irregularity in the structure. Chem Phys Chem 6:1442–1459

Kiri P, Hyett G, Binions R (2010) Solid state thermochromic materials. Adv Mat Lett 1:86–105

Kirk-Othmer (2013) Chemical technology of cosmetics. Wiley, Hoboken

Klán P, Wirz J (2009) Photochemistry of organic compounds. Wiley, Chichester, West Sussex

Lampert CM (1995) Chromogenic switchable glazing: towards the development of the smart window. In: Window innovations'95, Toronto

Lampert CM (2004) Chromogenic smart materials. Mater Today 7:28–35

Larson RG (1999) The structure and rheology of complex fluids. Oxford University Press, New York

Lee D (2007) Nature's palette: the science of plant color. The University of Chicago Press, Chicago

Lenau T, Barfoed M (2008) Colours and metallic sheen in beetle shells: a biomimetic search for material structuring principles causing light interference. Adv Eng Mater 10:299–314

Liu J, Coleman JP (2000) Nanostructured metal oxides for printed electrochromic displays. Mater Sci Eng A286:144–148

Lutz A (2007) Stress and/or temperature indicating composition for roll covers. US Patent 7,261,680 B2, 28 August 2007

Meunier L, Kelly FM, Cochrane C, Koncar V (2011) Flexible displays for smart clothing: part II – electrochromic displays. Ind J Fib Text Res 36: 429–435

Minuto A, Vyas D, Poelman W, Nijholt A (2012) Smart materials interfaces: a vision. In: Camurri A et al (eds) INTETAIN 2011, LNICST 78, Institute for Computer Sciences, Social Informatics and Telecommunications Engineering. Springer, Heidelberg, pp 57–62

Nandu S (2012) Ice-Ice Safety! Testing the effects of thermochromic signs on driver behavior. Google science fair. https://sites.google.com/a/googlesciencefair.com/science-fair-2012-project-b98f01851513e40127372d65820f09671f5eb48d-1332432587-88/home. Accessed 1 July 2013

Nelson G (2002) Application of microencapsulation in textiles. Int J Pharmaceuticals 242:55–62

New Scientist (1977) Anyone for technological tennis? 6 Oct 1977, p 26

Özgürler Trafik (2013) Termodinamik reflector (thermodynamic reflector) product leaflet and private communication

Patil PS (2000) Gas-chromism in ultrasonic spray pyrolyzed tungsten oxide thin films. Bull Mater Sci 23:309–312

Pawlicka A (2009) Development of electrochromic devices. Recent Pat Nanotechnol 3:177–181

Perry JD, Freydière AM (2007) The application of chromogenic media in clinical microbiology. J Appl Microbiol 103:2046–2055

Potisek SL, Davis DA, Sottos NR et al (2007) Mechanophore-linked addition polymers. J Am Chem Soc 129:12808–13809

Potisek SL, Davis DA, White SR et al (2010) Self-assessing mechanochromic materials. US Patent US 2010/0206088 A1, 19 August 2010

PPG Industries (2013) PPG Aerospace shows darker electrochromic window shade option at NBAA. http://www.ppg.com/en/newsroom/news/Pages/20101017A.aspx. Accessed 24 July 2013

Project TeTRInno SmarTex (2007) State of the art in smart textiles and interactive fa brics. http://www.mateo.ntc.zcu.cz/doc/State.doc. Accessed 15 Oct 2012

Rassart M, Colomer JF, Tabarrant T, Vigneron JP (2008) Diffractive hygrochromic effect in the cuticle of the Hercules beetle *Dynastes hercules*. New J Phys 10:1–14

Rauf MA, Hisaindee S (2013) Studies on solvatochromic behavior of dyes using spectral techniques. J Molec Struct 1042:45–56

Reichardt C (1994) Solvatochromic dyes as solvent polarity indicators. Chem Rev 94:2319–2358

Remes B (2009) SAGE Electrochromics, Inc. energy conservation. J Minn Precision Manuf Assoc IntrinXec Manage Inc., Minneap. http://www.mpma.com/pdfs/Journals/PM_May_June_09.pdf. Accessed 10 May 2013

Ritter A (2007) Smart materials in architecture, interior architecture, and design. Birkhäuser, Basel

Sage Glass (2013) http://en.wikipedia.org/wiki/SAGE_Electrochromics. Accessed 15 May 2013

Savill DG (2002) Toothbrush. US Patent US 6,389,636, 21 May 2002

Scindia Y, Silbert L, Volinsky R, Kolusheva S, Jelinek R (2007) Colorimetric detection and fingerprinting of bacteria by glass-supported lipid/polydiacetylene films. Langmuir 23:4682–4687

Seeboth A, Loetzsch D, Ruhmann R (2011) Piezochromic polymer materials displaying pressure changes in bar-ranges. Am J Mater Sci 1(2):139–142

Seeboth A, Ruhmann R, Mühling O (2010) Thermotropic and thermochromic polymer based materials for adaptive solar control. Materials 3:5143–5168

Simon RA, Nilsson KPR (2010) Optical reporting based on FRET between conjugated polyelectrolyte and organic dye for biosensor applications. In: Demchenko AP (ed) Advanced fluorescence reporters in chemistry and biology II. Springer, Heidelberg

Somani PR, Radhakrishnan S (2002) Electrochromic materials and devices: present and future. Mater Chem Phys 77:117–133

Song J, Cheng Q, Zhu S, Stevens RC (2002) "Smart" materials for biosensing devices: cell-mimicking supramolecular assemblies and colorimetric detection of pathogenic agents. Biomed Microdevices 4:213–221

Sonmez G et al (2004) A red, green and blue (RGB) polymeric electrochromic device (PECD): the dawning of the PEC

Su Y-L, Li J-R, Jang L, Cao J (2005) Biosensor signal amplification of vesicles functionalized with glycolipid for colorimetric detection of Escherichia coli. J Colloid Interface Sci 284:114–119

Stevens M, Merilaita S (2011) Animal camouflage: mechanism and function. Cambridge University Press, Cambridge

Suntek (2012) http://suntekllp.com/18.php. Accessed 1 Dec 2012

Takagi HD, Noda K, Itoh S (2004) Piezochromism and related phenomena exhibited by palladium complexes. Platinum Metals Rev 48:117–124

Talvenmaa P (2006) Introduction to chromic materials. In: Mattila H (ed) Intelligent textiles and clothing. Woodhead Publishing Ltd, Cambridge, pp 193–204

Tian et al (2003) A single photochromic molecular switch with four optical outputs probing four inputs. Adv Mater 15:2104–2107

Tiwari A et al (2011) Smart polymeric nanofibers resolving biorecognition issues. In: Li S et al (eds) Biosensor materials. Wiley, Weinheim

Tomasuolo M, Raymo F (2010) Use of oxazine compounds for making chromogenic materials. US Patent 0249403 A1, Sep 2010

USTA (2008) Instant replay makes its US open series debut. http://www.usta.com/Active/News/USTA-News/Press-Releases/336734_Instant_Replay_Makes_Its_US_Open_Series_Debut/. Accessed 29 Jan 2013

Weder C (2011) Mechanoresponsive materials. J Mater Chem 21:8235–8236

Van der Schueren L, De Clerck K (2013) Halochromic textile materials as innovative pH-sensors. In: Vincenzini P, Carfagna C (eds) Advanced in science and technology (Proceeding of 4th international conference on smart materials, structures and systems, Tuscany), vol. 80. Trans Tech Publications, Switzerland

Van Gemert B (1999) Organic photochromic and thermochromic compounds. In: Crano JC, Guglielmetti RJ (eds) Main photochromic families, vol 1. Plenum Press, New York, pp 111–140

Velusamy V, Arshak K, Korostynska O, Oliwa K, Adley C (2010) An overview of foodborne pathogen detection: in the perspective of biosensors. Biotechnol Adv 28:232–254

Vigneron JP, Colomer J-F, Vigneron N, Lousse V (2005) Natural layer-by-layer photonic structure in the squamae of *Hoplia coerulea* (Coleoptera). Phys Rev E 72:1–6

West K (2009) Animal behavior: animal courtship. Infobase, New York

Worbin L (2010) Designing dynamic textile patterns. Chalmers University of Technology, Dissertation

Zhong W (2013) An introduction to healthcare and medical textiles. DEStech, Lancaster

Chapter 3
Manufacturing and Processes Related to Chromogenic Materials and Applications

Abstract This chapter aims to explain various manufacturing processes used for commercial products that involve chromogenic materials. Challenges in the use of chromogenic pigments, inks, and dyes are presented based on industrial practice and patents. Some of the manufacturing processes are applicable to more than one type of chromogenic material type (thermochromic, photochromic, electrochromic, etc.) so the emphasis is on the process and the type of colorant (liquid crystal, leuco dye, inorganic oxide) in this chapter, rather than the specific chromogenic category. One of the difficulties in the use of chromogenic pigments or dyes is their incorporation to a bulk material or to the surface without harming the chromogenic features. Therefore, microencapsulation is a key process, allowing more flexibility in manufacturing and product use. The appropriate printing technique needs to be selected for different substrates and conditions of use. Some of the processes are explained with reference to the targeted product, including chromogenic textiles, gels, windows, and displays.

Keywords Liquid crystal · Leuco dye · Microencapsulation · Chromogenic ink · Textile · Display

3.1 Liquid Crystals and their Use in Temperature Indicators

Liquid crystals represent a special class of materials with characteristics of both liquid and solid phases. This unusual state of matter owes its unusual properties to orientational and sometimes partial positional order (Collings and Hird 1997; Larson 1999). Although liquid crystals possess some order, the physical properties are more liquid-like. Usually, they have the consistency between that of thin oil such as olive oil and viscous paste such as grease (Hallcrest 1991). Some soaps, phospholipids, and oils are liquid crystals. Considering their chemical structure, they can be amphiphilic molecules (molecules with a polar, water soluble group attached to a

M. Ferrara and M. Bengisu, *Materials that Change Color*, PoliMI SpringerBriefs, DOI: 10.1007/978-3-319-00290-3_3, © The Author(s) 2014

nonpolar, water-insoluble group) or polymers. Therefore, the synthesis of liquid crystal molecules can be achieved by common procedures used for the relevant chemical structure. However, liquid crystal molecules are typically unsuitable for applications in their raw form due to several reasons. Unprotected liquid crystal droplets are susceptible to degradation and contamination. They are sensitive to some organic chemicals such as fats, greases, and solvents. Furthermore, they tend to fuse, split, or smear on coated or printed films used for various applications (Hallcrest 1991; Sage 2011). Microencapsulation is a process of encasing individual liquid crystal droplets with a very thin, relatively durable layer. The droplet sizes can range from nanometers to millimeters. Encapsulation protects liquid crystals from external chemical and physical effects. Furthermore, it allows mixing of different liquid crystal formulations to adjust color and transformation temperatures. It also allows the deformation of droplets into oblate ellipsoids or flat rods which is necessary to achieve sufficient contrast in display applications (Sage 2011). Some microencapsulation processes are described in Sect. 3.3.

Typically, microencapsulated liquid crystals are used in the form of an ink. The ink can remain in the liquid state as part of the product. This approach is used in electrochromic displays. Chromogenic inks can also be applied on various substrates by printing, spraying, or painting, and then dried, as in the case of temperature indicators and special textiles. Liquid crystal containing microcapsules are currently not incorporated in high-temperature and high-pressure processes such as polymer extrusion because the microcapsules have limited processing stability (Seeboth and Lötzsch 2008). However, more rigid microcapsules or other strategies may be developed in the future to increase the range of processes and applications of liquid crystals.

Temperature indicators containing liquid crystals are cheap but dependable devices which can be designed and tuned for different temperature ranges, depending on the specific application. For example, a freezer thermometer requires a color play range of −30–10 °C, a beverage ideal temperature indicator used for beers, vines, or soft drinks requires a range of 5–15 °C, and a forehead thermometer requires a range of 35–40 °C. The manufacturing process for such products involves deposition of suitable inks to the desired areas by coating or printing. Coating can be achieved by bar or blade coating. Printing is usually achieved by screen, inkjet, or flexographic printing. The ink is composed of liquid crystal droplets and an appropriate binder. Water-soluble polymers such as poly(vinyl alcohol) as well as acrylic– or polyurethane-based emulsions are suitable binders. Low viscosity inks are suitable for inkjet or flexographic printing while thick paste formulations are necessary for screen printing. The droplet size should remain constant and be protected from contamination during the process. The droplets are deformed into oblate ellipsoids or flattened against a planar surface to align the liquid crystals to improve brightness. Oblate ellipsoids can be obtained upon drying a well-formulated ink. As mentioned before, microencapsulation is the preferred method to protect the liquid crystal droplets during these processes.

A commercial spray powder coating with the brand name Termocromico by Adapta Color S.L. uses microencapsulated liquid crystal emulsion mixed with

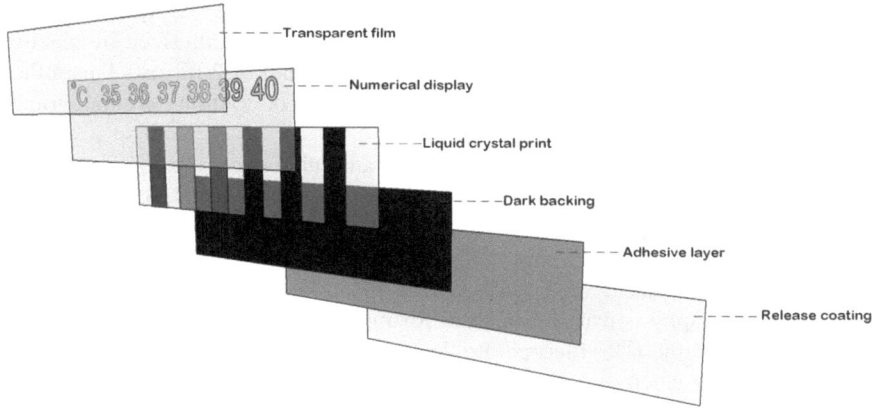

Fig. 3.1 Layers of a liquid crystal strip thermometer

polyurethane, UV inhibitors, binders, and fixers. This can be used as a durable thermochromic coating for accessories, apparel, handbags, upholstery, and the like (Material Connexion 2013).

A common form of liquid crystal temperature indicators is the strip thermometer. This thermometer is made up of several layers (Fig. 3.1). A black backing is typically used in indicators to maximize contrast. Liquid crystal inks are deposited on top of the black backing. White graphic print is used to indicate numbers or symbols. Several film layers having a thickness from 10–200 μm are sandwiched and laminated. Alternatively, the black backing layer can be printed onto the dried transparent plastic substrate which contains the liquid crystal ink (Sage 2011; Hallcrest 2012).

3.2 Leuco Dyes and their Incorporation into Polymers

Leuco dyes are the most important substances to make inks, paints, polymeric films, components or bulk products with chromogenic properties. Various types of chromogenic phenomena have been detected in different types of leuco dyes including thermochromism, photochromism, and electrochromism (Muthyala 1997). For example, three commercially important photochromic leuco dye groups are spyropyrans, spirooxazines, and naphthopyrans (also called chromenes). Indolinobenzopyrans belong to the group of spyropyrans. A classical example is indolinospyrobenzopyran. This molecule can be synthesized by the condensation of Fischer's base and salicilaldehyde in anhydrous ethanol or benzene as a solvent (Nakazumi 1997; Corns 2009). The isolated reaction product may be pure enough for certain applications. For higher purity, recrystallization or column chromatography is necessary.

Two important spirooxazine subclasses are naphth[2, 1-*b*; 1, 4]oxazines and naphth[1, 2-*b*; 1, 4]oxazines. These molecules are usually synthesized by reaction of a Fischer's base derivative with a 1-nitroso-2-naphthol or 2-nitroso-1-naphthol, in alcohol as a solvent. Purification by recrystallization or column chromatography is necessary.

Two commercially important naphthopyran subclasses are 3*H*-naphtho[2, 1-*b*]pyrans and 2*H*-naphtho[1, 2-*b*]pyrans. A general process for large-scale manufacture involves reaction of 1- or 2-naphthol derivatives with a diaryl alkynol compound. A non-polar solvent, such as toluene, is used in the presence of a suitable acidic substance, such as Al_2O_3. Purification by recrystallization or column chromatography is usually required for industrial applications (Corns 2009).

A suitable method to incorporate leuco dyes into various polymers is to microencapsulate leuco dye-developer-solvent systems and use the microcapsules as pigments during shaping the polymer in the plastic or liquid state. Microencapsulation allows the leuco dye system to act independently from the polymer matrix. Seeboth and his team at the Fraunhofer Institute use this approach to obtain thermochromic polymers in the form of foils, injection molded parts, extruded parts, and thermosetting parts (Seeboth and Lötzsch 2008; Fraunhofer 2012). Within the dye-developer-solvent system, leuco (colorless, from the Greek word *leukos* meaning white) dyes play the role of electron donating substances. The electron given by the dye is accepted by the developer, which donates a proton to the leuco dye. The developer is a weak acid that acts as a co-solvent. The role of the solvent is to act as a host to the dye and the developer while it should also enable the thermochromic reaction to be reversible with temperature. The solvent is polar and typically has a low melting point (Mills 2009). A good example to such systems is a mixture of 1 wt% 2-chloro-6-diethylamino-3-methylfluoran (leuco dye), 5 wt% 2, 2'-bis(4-hydroxyphenyl)-propane (developer) and 94 wt% 1-hexadecanol (solvent). When this system is cooled below 48 °C, the colorless liquid crystallizes and develops a vermillion color (Seeboth and Lötzsch 2008). Another such system is made up of crystal violet lactone (CVL) as the leuco dye, lauryl gallate as the color developer, and 1-tetradecanol or another long chain alcohol as the solvent. A typical molar ratio for these components is 1:6:40, respectively (Seeboth et al. 2010). Figure 3.2 shows a general scheme for the coloration and discoloration process in dye-developer-solvent systems.

3.3 Microencapsulation

Various methods exist for microencapsulation of liquid crystal droplets and other types of dyes including coacervation, spray drying, interfacial polymerization, and in situ polymerization. Coacervation is the most frequently used method for industrial applications. Coacervation processes are classified as simple or complex. The first commercially significant microencapsulation process was a complex coacervation process used for carbonless copy paper developed by NCR

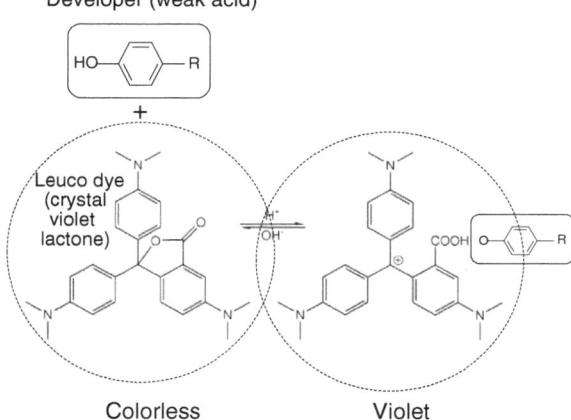

Fig. 3.2 An example of coloration and discoloration in the leuco dye-developer-solvent system

Corporation in the 1950s (Hatano 1997; Hallcrest 1991). This process involves the following steps (Fig. 3.3):

- Preparation of an emulsion of a leuco dye and aqueous gelatin solution.
- Addition of gum arabic solution.
- Coacervation by adjusting the pH with acetic acid. At this step, gelatin has a positive surface charge due to the acidic nature of the solution while gum arabic has a negative charge regardless of pH.
- Cooling below 10 °C, resulting in gelation of the liquid polymer (gelatin) on the leuco dye core.
- Hardening of the gelled coating with formaldehyde.

All these steps take place while stirring the solution. A similar process can be used for the microencapsulation of liquid crystals. For example, thermochromic liquid crystals can be encapsulated using pigskin gelatin, gum acacia (same as gum arabic), and formaldehyde or glutaraldehyde as the cross-linking (hardening) agent (Sage 2011).

Two widely used techniques for microencapsulation of thermochromic and photochromic dyes are based on urea and melamine–formaldehyde systems (Nelson 2001). The melamine–formaldehyde system uses a mixture of water, dye, oil, and melamine formaldehyde. The mixture is vigorously shaken to create a very fine emulsion. After the chemical reaction (or setting), a polymer containing a dispersion of encapsulated chromogenic dye is obtained, which is very hard and strong. Melamine formaldehyde creates the capsules which contain the oil and dye (Small and Highberger 2000).

Although microencapsulation protects chromogenic dyes to a certain extent, heat, UV light, pressure, and solvents can degrade their quality. For example, microencapsulated thermochromic dyes applied on textiles can survive up to 20

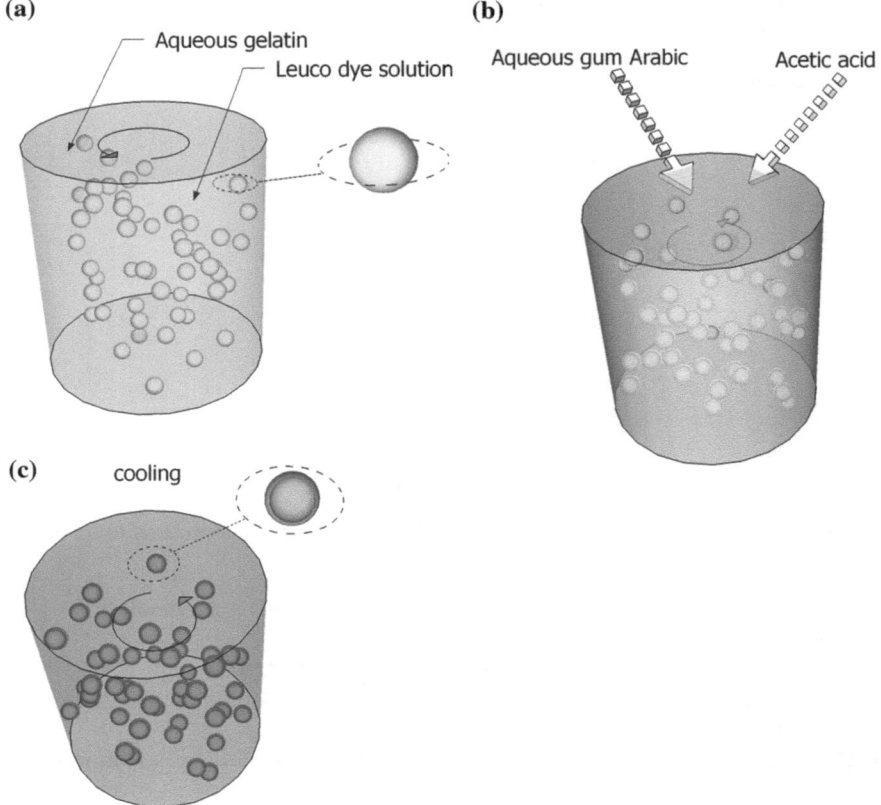

Fig. 3.3 Microencapsulation of chromogenic dye by complex coacervation. **a** Emulsion, **b** Coacervation, **c** Gelation

laundering cycles but drying at elevated temperatures or use of bleach can reduce the durability (Nelson 2001).

Typically, microcapsules have a particle size of 1–10 μm for leuco dyes and 5–150 μm for liquid crystals. For instance, the particle size of microcapsules containing liquid crystal molecules prepared by the emulsion method was reported to range from 4 to 10 μm (Chen et al. 2009). Commercially available forms of microencapsulated liquid crystals include adhesive sheets, ready-to-spray slurries, and water-resistant microcapsules (Smith et al. 2001).

3.4 Chromogenic Inks

Two types of chromogenic inks are those containing leuco dyes and liquid crystals. Commercially available chromogenic inks include thermochromic, photochromic, and electrochromic colorants. Colorants used in inks and other media are either pigments or dyes (Leiby 2004). Simply put, a pigment is considered to be insoluble in a liquid or solid medium. Pigments are typically molecular or crystalline particles which remain unaffected by the surrounding medium. On the other hand, dyes are soluble in the liquid or the solid (for example polymer) host. In spite of the technical definition given here, the distinction between pigments and dyes is not always clear-cut.

Since pigments remain as a separate phase in the host medium, they normally reduce transparency due to light diffraction while it is possible to obtain water-clear (i.e. optical) transparency with dyes. Furthermore, dyes provide strong, bright colors while pigments create a weaker effect. The disadvantages of dyes are their lower heat stability and their migration in certain polymers such as PVC and polyolefins. The use of pigments as colorants requires special considerations in manufacturing due to the relative difficulty in wetting and dispersion. A common industrial solution to this challenge is to provide the pigments in a concentrated form highly dispersed in a low molecular weight polymeric matrix. Another possibility is to provide easy dispersing pigments with fine particle size (Christensen 2003).

Inks are composed of pigments or dyes dispersed in suitable vehicles. Ink vehicles can be resins dissolved in solvents or oils. Dispersants are added to improve dispersion of the colorants while dryers are utilized to aid drying of the ink. The ink formulation changes according to the type of printing method. For example, inks used for flexography and gravure consist of colorants, resins, and suitable solvents including water, alcohol, or other organic solvents (Leach and Pierce 2007).

The following are some examples giving details of special chromogenic ink formulations from the patent literature.

Example 1 Thermochromic quick set lithographic ink (Small and Highberger 2000)

Ingredient	Weight (%)
Offset ink base	75.0
Quick set gel vehicle	12.5
Quick set free flow vehicle	7.5
12 % cobalt drier	1.0
6 % manganese drier	1.0
Ink oil (initial boiling point 265 °C)	3.0

Example 2 Thermochromic ink for security document (Mehta et al. 2002)

Ingredient	Weight (%)
Phenolic modified resin	35.6
Tall oil fatty acid ester	31.8
Thermochromic aqueous slurry	32.6

Example 3 Photochromic yellow ink for ink-jet printing (Meinhardt and Bridgeman 2000)

Ingredient	Weight (%)
Photochromic yellow dye	0.5–15.0
Urethane or urethane/urea resin	15.0–50.0
Mono-amide	30.0–70.0

Thermochromic leuco dye inks have a shelf life of six months or more while stabilized photochromic inks can last for years before printing. After printing, thermochromic leuco dye inks can last for years if not exposed to too much UV-light or temperatures over 140 °C. Photochromic inks can only last up to a few months of daylight exposure after printing. Thermochromic and photochromic textile inks have the ability to withstand about 20 wash cycles and they are both vulnerable to bleach (Talvenmaa 2006, Kulčar 2012).

Various types of chromogenic inks are available commercially. Solar Zone (Liquid Crystal Resources) and Dynacolor (Chromatic Technologies) are photochromic inks which are colorless indoors and become brightly colored when exposed to sunlight. These inks are suitable for various printing processes such as offset, screen, and flexographic printing. Chromicolor (Matsui) is available as a microencapsulated leuco dye ink. This water-based thermochromic ink changes from a colorless state or one color to another at a desired temperature (Material Connexion 2013).

3.5 Chromogenic Gels

A *hydrogel* is a macromolecular network formed of hydrophilic polymers swollen with water or aqueous liquids (Peppas 1987). Various approaches have been studied to prepare thermochromic hydrogels. One approach uses the interaction of pH-sensitive indicator dye molecules with the microenvironment provided by the hydrogel. Two types of such dyes, DTTP (Reichardt's Dye) and Cresol Red, have been embedded into an aqueous polyvinyl alcohol (PVA)-borax-surfactant hydrogel. The DTTP-embedded hydrogel is colorless at pH 8.5 and 10 °C. Upon heating, the color changes reversibly and gradually to a deep violet at 80 °C. The Cresol Red-aqueous PVA system is an optically clear yellow gel at room

(a)   (b)   (c)

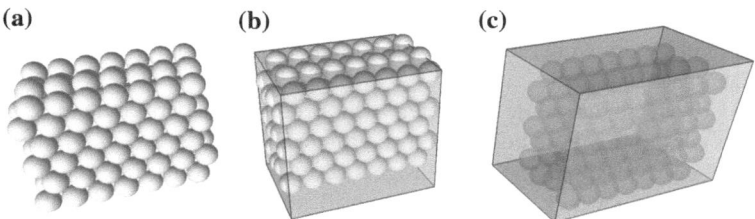

Fig. 3.4 Preparation of a gel matrix with periodically ordered pores: **a** Close packed silica colloidal crystal, **b** Infiltration of the silica template with the pre-gel solution and polymerization to form the gel, **c** Periodically ordered porous gel obtained by removal of silica particles by immersion into HF solution

temperature and pH 8.5. Upon heating, its color changes reversibly to wine-red at 80 °C (Seeboth 1999).

Thermochromic gels can also be obtained by the self-assembly of uniform, highly charged colloidal particles in liquid media of very low ionic strength into a crystalline array. The crystallinity of the array results in Bragg reflection (reflection of radiation by crystalline or periodic systems). Thermal changes result in swelling or shrinking of the crystalline array and in turn, cause a variation in Bragg reflection and an accompanying change in color. This concept was demonstrated with highly charged polystyrene colloidal spherical particles with a uniform diameter of 99 nm. The particles were dispersed in N-isopropylacrylamide solution in which they formed a body centered cubic array of swollen spheres. Changing the temperature from roughly 12–35 °C resulted in a continuous change of the diffracted light wavelength from 704 to 460 nm, which corresponds to color variation from red to blue (Seeboth and Lötzsch 2008). These three-dimensionally periodic structures with a periodicity scale comparable to the wavelength of light have been coined *soft photonic crystals*. Ueno et al. (2007) used a different approach to prepare photonic crystals in the gel form. Instead of the opal type soft photonic crystal which is comprised of periodically ordered colloidal particles, they prepared a gel matrix with periodically ordered pores (Fig. 3.4). In order to obtain such a structure, they first prepared a close packed silica colloidal crystal by the self assembly of silica spheres 210 nm in diameter from a suspension of silica gel particles. This silica colloidal crystal was then dried and used as a template to create a gel. The pre-gel solution was infiltrated into the voids of the template and it was polymerized to obtain the gel. Methacrylic acid (MAAc) as a pH-sensitive monomer, N-isopropylacrylamide (NIPA) as a thermosensitive monomer and N, N'-methylene-bis(acrylamide) (BIS) as a cross-linker were used to obtain the hydrogel. The silica particles inside the gel were dissolved by immersing the whole mass into 5 wt% HF aqueous solution for one week. This process resulted in a porous gel membrane with a thickness of ~ 0.5 mm. Such porous gels are able to recall their three dimensional shape due to the presence of cross-linkers used during polymerization. The color of such a soft photonic crystal can be tuned due to swelling and shrinking of the gel by various stimuli including pH, temperature, and UV light (Harun-Ur-Rashid et al. 2010). Such soft photonic crystals are a new generation of chromogenic materials which may have very interesting applications in the near future.

Leuco dyes are insoluble in water, which creates a barrier against forming hydrogels containing these colorants. Nevertheless, Babic et al. (2009) managed to prepare leuco dyes dissolved in hydrogels by using micelles (aggregates of surfactant molecules) which provide the hybrid polar-nonpolar environment. Leucomalachite green (LMG) and leuco crystal violet (LCV) dyes were dissolved in 4 % gelatin hydrogels with Triton X-100 surfactant. The gelatin solutions were allowed to gel overnight in a refrigerator at 5 °C. The colorless hydrogels were used to demonstrate their potential as three dimensional radiochromic dosimeters for optical computed tomography scanners. Irradiation of these gels with radioactive cobalt-60 beams resulted in stable color images of the dose. While this study involved radiation therapy 3-D dose verification, such gels could also be developed for other thermochromic or photochromic applications.

Characteristics of thermotropic gels were discussed in this chapter. An example to phase separating thermotropic gels is Cloud Gel. This gel is prepared by cross linking of an aqueous poly(methyl vinyl ether) solution with methylene-bis-acrylamide (Seeboth et al. 2010). The switching temperature can be set between 10 and 70 °C by altering the concentration of the components (Beck et al. 1995).

3.6 Printing and Coating with Chromogenic Inks and Dyes

Various printing technologies are available for different industries, which can be used to apply chromogenic inks on different types of surfaces including paper, fabric, metal, and polymer film. However, the special nature of chromogenic inks may require some precautions or adaptations. For example, chromogenic molecules can degrade if high temperatures are applied during drying or any other stage of the process. Commercial photochromic dyes have been reported to degrade above 240 °C after a few minutes while thermochromic leuco dye inks degrade above 140 °C (Nelson 2001; Kulčar 2012).

In *roller printing*, designs or patterns are transferred onto the substrate via engraved printing rollers. This is a continuous process with high print rates. Sharply defined and brilliant patterns are possible by this technique although investment and manufacturing costs are high. *Screen printing* is a versatile process with many applications and variants. Other terms used for this process are serigraphy and stencil printing. The forms to be printed are transferred onto the textile, paper or other material through stencils which contain spots which are permeable to the ink (Fig. 3.5). The rest of the stencil is impermeable to the ink. The patterns are applied through a photochemical process onto a fine gauze sieve covered with a layer of photosensitive polymer. The stencil is applied to a mesh, both of which make the screen. Commonly, a potent, prevalently UV light source, is applied to the stencil through a negative film. Parts of the stencil exposed to light remain soft

Fig. 3.5 Screen printing
basics: **a** Ink **b** Squeegee
c Image **d** Photo-emulsion
e Screen **f** Printed image
(Image: Harry Wad,
Wikimedia Commons)

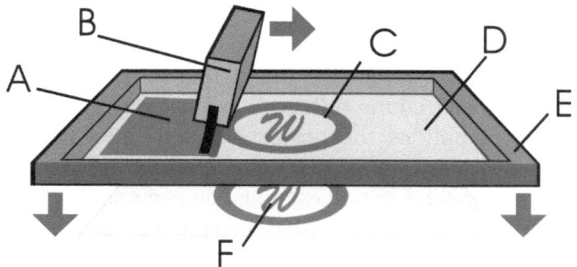

and are washed by water, leaving suitable openings for printing. The rest of the stencil becomes hard and impermeable. However, this strong UV light may be detrimental to certain chromogenic dyes. Certain alternatives exist to photosensitive emulsions. The simplest approach is to prepare the desired image by cutting out a stencil from paper, cardboard, or sheet material that can be easily cut. This stencil is attached to the silk screen and the printing ink is applied through them onto the fabric using a squeegee. An industrially more dependable solution might be to use thermal imaging. Images or photos can be etched or burned onto heat sensitive films using thermal imagers, eliminating the need for emulsions or UV exposure. The resulting stencil or film is attached to a plastic frame for screen printing (Laury 1997).

Multicolor screen printing requires a different stencil for each color. Screen printing applied by a flat stencil is called rectilinear stencil printing or flat bed printing. *Rotary stencil printing* uses hollow cylindrical stencils. The rotary action allows continuous printing and provides higher print speeds (Giles 2000; Wulfhorst et al. 2006).

Flexography is a high quality printing process applied on a wide range of materials such as flexible packaging, corrugated board, or cardboard. In this process, mainly three rolls are used. The anilox roll transfers a controlled amount of ink to the printing plate. The printing plate contains the raised image on a surface of flexible material such as rubber or polyester, secured on a cylinder. During the printing process, the easy-drying ink is transferred from one cylinder to the other and finally onto the material to be printed on. The substrate, for example a packaging film, passes through the printing cylinder which prints the desired image on the substrate, and an impression cylinder which applies pressure (Giles 2000).

Offset printing is one of the principal planographic printing processes where the positive and negative images are on one flat plane. The process is indirect where the image is transferred from the inked image to an intermediate cylinder covered with a rubber blanket and from this to the printing substrate. This process is the most diffused printing process thanks to its simplicity, low cost, durability, reliability, and versatility with regard to machine's and types of substrates to be printed. The printing plates can be paper, plastic or metal. The image areas accept ink while they repel water. In contrast, the negative image areas reject ink while they attract water. Offset printing machines can be of foil or bobbin type, single or multicolor, with a printing plate on a single side or on both sides. Their structure is composed of three printing cylinders coated with rubber and two groups of washing and inking rollers. Waterless offset printing is an offset printing process which simplifies the process through the elimination of the dampening step. During waterless offset printing, the ink is applied to a dry cylinder, usually made from aluminum or polyester and coated with a thin layer of silicone. The repelling force between special types of silicones and inks is suitable for the non-image areas. The image areas are made from a photopolymer (Romano 1999) .

Coating is an operation where a chemical substance with suitable rheology is spread on a continuous substrate such as textile, paper, or polymer. The purpose of coating is to impart new characteristics to the substrate or to protect it from environmental effects. One or both sides of the substrate can be coated using processes such as rotary stencil coating, reverse roll coating, slot die coating, gun spraying, and dip coating. The coating material can be applied directly or indirectly. The indirect method uses an intermediate substrate such as film or paper, which can eventually be removed (Wulfhorst et al. 2006). Chromogenic pigments can be added to the coating solution and applied to the surfaces of textiles or other substrates. For example, a polymer coating with thermochromic pigments can be applied on clothing as a visual indicator of thermal changes in the environment (Stylios 2006).

3.7 Chromogenic Textiles

Commonly used techniques for incorporating chromogenic pigments into textile include printing, coating, and dyeing (Hu 2010). Microencapsulated pigments are preferred in printing and coating processes but not in dying. In the textile industry, *printing* involves the transfer of designs onto the fabric. Currently, the most important printing processes for textiles are roller printing, screen printing, and rotary stencil printing (Wulfhorst et al. 2006), shortly explained in the previous section.

Dyeing is the process of imparting color to fibers using natural or synthetic colorants to make them permanently colored (Chatwal 2009; Smith 2009). It can be applied to various forms of fibers including individual fibers, yarn, or fabric. Fibers and their assemblies can be dyed thanks to a physical or chemical affinity between the dye and the fiber. Disperse or dispersion dyes are dyes which dissolve in the fibrous material. Important process parameters in dyeing include the composition and temperature of the dyeing liquor, dyeing time, pH, and the rate of fabric. Various industrial processes exist for dyeing, including spin, flock, top, yarn, and piece dyeing (Wulfhorst et al. 2006). *Coating*, in textile terminology, is the process of applying a thin layer (4–50 g/m^2) of chemical finishing with the purpose of imparting a desired property onto the surface, not present in the substrate (Schindler and Hauser 2004).

The development of specific techniques suitable for chromogenic colorants still continues since commercialization of chromogenic technologies is relatively new. A series of studies on textile applications of photochromic dyes was conducted at the School of Textiles and Design, Heriot-Watt University (Little and Christie 2010). The use of commercial photochromic dyes in screen-printing was analyzed. Two types of screen printing methods used in this study were pigment printing and printing with disperse dyes. The pigment screen print formulation was prepared from commercial photochromic dyes (naphthooxazines or naphthopyrans), Acramin binder, Magnaprint Clear M04, emulsifier BR, Acramin Softener S, Acrafix ML, and water. Note that although these dyes are not designed for textiles and they are not pigments but dyes (since they have some solubility) the pigment printing method worked well for them (Christie 2012). The disperse dye formulation contained photochromic dye, a pine-oil based low foam wetting agent, a dispersing agent (based on the disodium salt of a naphthalene sulphonic acid/formaldehyde condensate), a thickener (sodium alginate), and water. The paste was printed on polyester fabric, dried, and fixed by steam or dry heat at a temperature range of 160–180 for 5–10 min. Screen printing of disperse dyes and pigments was carried out using a plain mesh print screen on a magnetic print table. Printed samples were dried and fixed in an oven. When printed cotton and polyester fabrics were subjected to wash fastness test, the degree of photocoloration of naphthopyran dyes decreased progressively with washing, while it increased with initial washing in the case of the selected spirooxazine dyes, and then decreased with subsequent washing. The incorporation of UV absorbers and hindered amine light stabilizers (HALS) improved the photostability of the dyes. Spirooxazine dyes faded much more rapidly than naphthopyrans. However, all five commercial dyes investigated, which belong to naphthooxazine and naphthopyran classes, showed rapid color development upon UV irradiation and rapid fading characteristics upon removal of the UV source.

An alternative to printing, coating, or dyeing of textiles is to incorporate chromogenic pigments into molten polymers and produce fibers from the polymer by melt spinning or other fiber-making processes. This approach was used to make embroidery threads from molten polypropylene and different photochromic pigments. The thread appears white at the absence of UV-light and switches to a specific color under UV-exposure. The photochromic effect survived for more than two thousand cycles of UV-exposure. Fibers obtained by this method have the advantage that the pigments cannot be washed off during laundry since they are not present only at the surface but within the whole fiber (Talvenmaa 2006). Furthermore, a chromogenic effect can be applied to any part of a textile product, depending on where the chromogenic fibers are applied during manufacturing.

3.8 Photochromic Eyeglass Lenses

The manufacturing process of photochromic eyeglass lenses depends very much on whether the lens is made of glass or plastic. Corning scientists discovered silver halide based photochromic glass in 1964 (Van Gemert 2000). This glass is based on a borosilicate composition which contains a small amount (<1 %) of silver halide. After melting, fining, and conditioning the glass, automatic presses are used to produce several thousand blank spectacle lenses per hour (Corning 2012). One of the most critical stages of photochromic lens making is annealing. This is the stage where the blanks are heat treated at a temperature between the annealing strain point and the softening point, resulting in the formation of 4–15 nm sized silver halide crystals within the borosilicate glass matrix (Shelby 2005; Pulker 1999). The annealing temperature ranges between 550 and 700 °C according to composition. The blanks are slowly cooled at a controlled rate to room temperature. The annealing and cooling stages are closely regulated for optimum results. More recently developed photochromic compositions include silver molybdate, silver tungstate, copper halides, and cadmium halides, which have similar spectral sensitivities to silver halides but a stronger darkening effect at comparable light intensities (Pulker 1999). Silver halide glasses are still preferred to other photochromic materials because of their remarkable reversibility and extreme fatigue resistance to darkening-clearing cycles. However, glass lenses are heavy and much more brittle compared to plastic lenses. Furthermore, plastic lenses are easier to manufacture and thus, they are cheaper.

The casting technology developed by Corning for photochromic plastic lenses uses a transparent liquid monomer and different types of photochromic dyes. Corning provides this resin mixture to lens manufacturers worldwide, which use the recommended lens casting procedure. The typical casting process starts by mixing the resin mixture with a catalyst. This mixture is degassed and cast into molds in a clean room. It is then cured by UV and heat or by heat alone. The cured lens is removed from the mold, cleaned, rinsed, polished, and coated with a 3–5 μm thick protective layer. An anti-reflective coating can also be applied afterwards (Corning 2012).

3.9 Electrochromic Displays

Various materials have been considered and used for electrochromic displays. Among them, liquid crystals have been dominating the market with devices known as liquid crystal displays (LCDs). Such displays have found a wide range of applications ranging from small displays used in watches, calculators, interfaces for appliances, etc. to intermediate and large scale displays used especially in TVs, advertisement displays, and information boards (in airports, hospitals, train stations, and the like). It is noteworthy that newer generation technologies such as most LED TVs still use liquid crystal displays while the main difference is the backlighting (actually mostly positioned at the edges) provided by LED lighting instead of fluorescent backlighting. Nevertheless, true LED and OLED displays, are also available, which are emissive technologies where each pixel is its own light source.

Materials other than liquid crystals can also be used for the electrochromic displays. These include transition metal oxides (TMOs) also known as solid state electrochromics, Prussian blue and other polynuclear transition metal hexacyanometallates, phtalocyanines, viologens, buck minsterfullerene, and conducting polymers. Among these alternatives, conducting polymers seem to have the greatest potential for future markets because they offer significant advantages for designers, manufacturers, and end users. While TMOs pose challenges in preparation and processing, polymeric electrochromics are relatively easy to process, devices made with them are less costly, and their contrast and lifetime is better compared to TMOs. Among conducting polymers, currently the largest interest is devoted to those based on poly(aniline) (PANI), poly(pyrrole) (PPy), and poly(3,4-ethylenedioxythiophene) (PEDOT) (Somboonsub et al. 2010). PEDOT is the most popular one due to its high conductivity, ranging from 10^{-2} to 10^{5} S/cm (Somani and Radhakrishnan 2002).

Regardless of the electrochromic material used, a generic electrochromic display can be built using five or more layered battery type design. As depicted in Fig. 3.6, the electrochromic layer faces an electrolyte which acts as an ion conductor. These two layers (electrochromic and electrolyte) are sandwiched between two electrodes which must be made of conducting materials. One of these layers

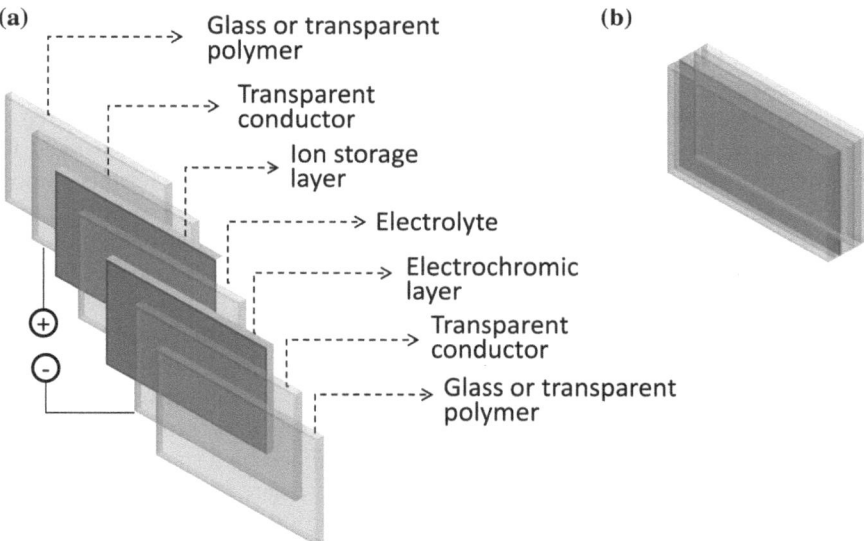

Fig. 3.6 a Principal layers of an electrochromic display **b** the assembled cell

should be transparent since it will function as the viewing side of the display. This is usually made of indium tin oxide (ITO) which satisfies the transparency and conductivity requirements. The counter electrode does not always have to be transparent, but it must satisfy a suitable redox reaction (Somani and Radhakrishnan 2002). The front of the display is covered with a glass layer to contain the electrode and to provide high optical quality. The back of the display may also have a glass layer which is either backed by a reflector, in the case of passive displays, or which is illuminated by a light source for active displays. In LCDs, in addition to the basic layers discussed so far, polarizing films on both sides of the layered package are used to polarize the light. Furthermore, spacers made of glass fibers or plastic balls are applied to assure the desired liquid crystal layer thickness. Since liquid crystals can easily flow, they need to be sealed with appropriate sealants (Collings and Hird 1997).

In flexible electrochromic displays, glass layers are replaced with a suitable plastic substrate such as PET and the electrodes are made of conductive polymers such as PEDOT (Meunier et al. 2011). Such flexible displays are being developed for textile and wearable device applications.

3.10 Smart Electrochromic Windows

The configuration and principle of operation of smart windows based on electrochromics is basically the same as electrochromic displays. The main difference is that both sides of smart windows should be transparent and there is no need for backlighting. Various technologies have been offered for smart electrochromic windows such as laminated battery type systems, suspended particle devices, and liquid crystal based devices. Here we will discuss the more common battery type design based on solid state electrochromics.

Unlike electrochromic displays where liquid crystal materials are preferred as the electrochromic material, the feasibility of TMOs has been demonstrated for smart window applications. The most common material employed for this purpose is WO_3, but alternatives such as MoO_3, V_2O_5, and Nb_2O_5 are also being investigated (Somani and Radhakrishnan 2002). Various firms have adopted the laminated battery type design for smart glass applications. Saint-Gobain (France), Flaberg/EControl Glass (Germany) and Gesimat (Germany) employed glass substrates. Gesimat used polyvinyl butyral (PVB) as the electrolyte, fluorine doped tin oxide (SnO_2:F) as the transparent ion conductor, and Prussian Blue as the ion storage material. The fluid electrolyte is injected in a gap between two glass substrates under vacuum in the case of Flabeg/EControl Glass' process but this is a slow process. Gesimat uses a PVB film which is applied by a standard lamination technique used for safety glass involving heat and pressure applied by rollers. The SnO_2:F coating is applied by spray pyrolysis during float glass production.

A different approach involves flexible PET foil as the transparent substrate. This low cost process, developed and commercialized by Chromogenics AB

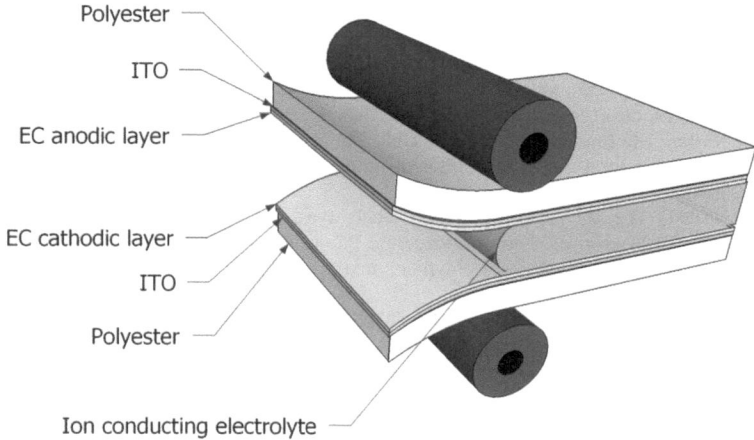

Polyester

ITO

EC anodic layer

EC cathodic layer

ITO

Polyester

Ion conducting electrolyte

Fig. 3.7 Roll forming process schematic depicting the multilayer structure of flexible ConverLight™ foil for smart window applications

(Sweden) uses roll to roll web coating where one PET foil is coated with ITO and NiO as the ion storage layer (Fig. 3.7). The two coated PET foils are then joined by an electrolyte at the middle, using continuous lamination (Granqvist 2012). ITO and electrochromic coating thicknesses in laminated smart glass systems are typically in the 100–300 nm range (Wittkopf 2010).

References

Babic S, Battista J, Jordan K (2009) Micelle hydrogels for three-dimensional dose verification. J Phys: Conf Ser 164:1–6

Beck A, Körner W, Scheller H, Fricke J, Platzer WJ, Wittwer V (1995) Control of solar insolation via thermochromic light-switching gels. Solar Energy Mater Solar Cells 36:339–347

Chatwal GR (2009) Synthetic dyes. Himalaya Publishing House, Mumbai

Chen KF, Wang F-H, Chan LH, Hsu SC, Chien YH, Liu SH, Cheng KL (2009) Multicolor polymer disperse microencapsulated liquid crystal displays. J Displ Technol 5:184–187

Christensen I (2003) Developments in colorants for plastics. Smithers Rapra, Shrewsbury

Collings PJ, Hird M (1997) Introduction to liquid crystals. Taylor & Francis, London

Corning (2012) High-index, photochromic, and sunglass blanks. http://www.corning.com/ophthalmic/products/glass_products/high_index_photochromic_sunglass_blanks.aspx. Accessed 13 Nov 2012

Corns SN, Partington SM, Towns AD (2009) Industrial organic photochromic dyes. Color Technol 125:249–261

Christie RM (2012) personal communication

Fraunhofer IAP (2012) Thermochromic polymers. http://www.thermochromic-polymers.com/thermochromic/index.html. Accessed 5 December 2012

Giles GA (ed) (2000) Design and technology of packaging decoration for the consumer market. Sheffield Academic Press, Sheffield

Granqvist CG (2012) Oxide electrochromics: An introduction to devices and materials. Sol Energy Mater Sol Cells 99:1–13

Hallcrest (1991) Handbook of thermochromic liquid crystal technology. http://www.hallcrest.com/downloads/RT006%20randtk_TLC_Handbook.pdf. Accessed 11 Oct 2012

Hallcrest (2012) Liquid crystals. http://www.colorchange.com/liquidcrystals. Accessed 11 Oct 2012

Harun-Ur-Rashid M, Imran AB, Seki T, Ishii M, Nakamura H, Takeoka Y (2010) Angle-Independent structural color. In: Colloidal amorphous array. Chemphyschem. doi:10.1002/cphc.200900869

Hatano Y (1997) The chemistry of fluoran leuco dyes. In: Muthyala R (ed) Chemistry and applications of leuco dyes. Kluwer Academic Publishers, Hingham, pp p159–p203

Hu J (2010) Adaptive and functional polymers, textiles and their applications. Imperial College Press, London

Kulčar R, Klanjšeg Gunde M, Knešaurek N (2012) Dynamic colour possibilities and functional properties of thermochromic printing inks. Acta Graphica 23:25–36

Larson RG (1999) The structure and rheology of complex fluids. Oxford University Press, New York

Laury JR (1997) Imagery on fabric. C&T Publishing, Lafayette

Leach R, Pierce RJ (eds) (2007) Printing ink manual. Springer, Dordrecht

Leiby FA (2004) Visual color matching in plastic materials. In: Charvat RA (ed) Coloring of plastics: fundamentals, vol.1. Wiley, Hoboken

Little AF, Christie RM (2010a) Textile applications of photochromic dyes. Part 1: establishment of a methodology for evaluation of photochromic textiles using traditional colour measurement instrumentation. Color Technol 126:157–163

Little AF, Christie RM (2010b) Textile applications of photochromic dyes. Part 2: factors effecting the photocoloration of textiles screen-printed with commercial photochromic dyes. Color Technol 126:164–170

Little AF, Christie RM (2010c) Textile applications of photochromic dyes. Part 3: factors affecting the technical performance of textiles screen-printed with commercial photochromic dyes. Color Technol 127:275–281

Material Connexion Database (2013) Sandow, New York. http://library.materialconnexion.com. Accessed 21 Feb 2013

Mehta R, Shields RL, Kalman AGL (2002) Thermochromic ink compositions. US Patent 6,413,305 B1, Jul 2002

Meinhardt MB, Bridgeman RR (2000) Photochromic ink. US Patent 6,022,909, Feb 2000

Meunier L, Kelly FM, Cochrane C et al (2011) Flexible displays for smart clothing: part II—electrochromic displays. Indian J Fibre Text Res 36:429–435

Mills A (2009) Intelligent inks in packaging. In: Yam KL (ed) Encyclopedia of packaging technology, 3rd edn. Wiley, Hoboken, pp p598–p605

Muthyala R (ed) (1997) Chemistry and applications of leuco dyes. Kluwer, Hingham

Nakazumi H (1997) Spiropyran leuco dyes. In: Muthyala R (ed) Chemistry and applications of leuco dyes. Kluwer, Hingham, pp 1–43

Nelson G (2001) Microencapsulation in textile finishing. Rev Prog Color 31:57–64

Peppas NA (1987) Hydrogels in medicine and pharmacy. CRC Press, Boca Raton

Pulker HK (1999) Coatings on glass. Elsevier, Amsterdam

Romano FJ (1999) Professional prepress, printing, and publishing. Prentice Hall, Upper Saddle River

Sage I (2011) Thermochromic liquid crystals. Liq Cryst 38:1551–1561

Schindler WD, Hauser PJ (2004) Chemical finishing in textiles. Woodhead Publishing, Cambridge

Seeboth A (1999) The first example of thermochromism of dyes embedded in transparent polymer gel networks. J Mater Chem 9:2277–2278

Seeboth A, Lötzsch D (2008) Thermochromic phenomena in polymers. Smithers Rapra, Shropshire

Seeboth A, Ruhmann R, Mühling O (2010) Thermotropic and thermochromic polymer based materials for adaptive solar control. Materials 3:5143–5168

Shelby JE (2005) Introduction to glass science and technology. Royal Society of Chemistry, Cambridge

Small LD, Highberger G (2000) Thermochromic ink formulations and methods of use. US Patent 6,139,779, Oct 2000

Smith CR, Sabatino DR, Praisner TJ (2001) Temperature sensing with thermochromic liquid crystals. Exp Fluids 30:190–201

Smith JL (2009) Textile processing. Abhisek Publications, Chandigarh

Somani PR, Radhakrishnan S (2002) Electrochromic materials and devices: present and future. Mater Chem Phys 77:117–133

Somboonsub B, Srisuwan S, Invernale MA et al (2010) Comparison of the thermally stable conducting polymers PEDOT, PANi, and PPy using sulfonated poly(imide) templates. Polymer 51:4472–4476

Stylios GK (2006) Engineering textile and clothing aesthetics using shape changing materials. In: Mattila H (ed) Intelligent textiles and clothing. Woodhead Publishing, Cambridge

Talvenmaa P (2006) Chromic and conductive materials. In: Mattila HR (ed) Intelligent textiles and clothing. Woodhead Publishing Limited, Cambridge

Ueno K, Matsubara K, Watanabe M, Takeoka Y (2007) An electro- and thermochromic hydrogel as a full-color indicator. Adv Mater 19:2807–2812

Van Gemert B (2000) The Commercialization of plastic photochromic lenses: a tribute to John Crano. Molecular crystals and liquid crystals science and technology. Sect A Mol Cryst Liq Cryst 344:57–62

Wittkopf H (2010) Elektrochrome Beschichtungen. ViP 22:26–30

Wulfhorst B, Gries T, Veit D (eds) (2006) Textile technology. Hanser Publishers, Munich

Chapter 4
Materials that Change Color for Intelligent Design

Abstract This chapter introduces key issues on the use of materials that change color as a design theme to experiment with its potential. It addresses issues such as the significance of color in design, design approaches toward and meaning attributed to dynamic color, theoretical implications and changes in design methodologies in the face of new tools and finally, technical and economic considerations useful for designers. Various design examples that address some of these key issues are presented.

Keywords Color design · Material design · Chromogenic materials · Design methods · Product design · Perception · Surface design · Textile design

4.1 Color and Design

What is the meaning of color for designers? What is the added value of color in the designed world? How should different colors be matched in a product, in fashion, in a home interior? There are many such questions trying to be answered by books and articles about design. This chapter aims to answer the first two questions very briefly and based on these answers, certain inferences will be made regarding how chromogenic materials could change current practices on the use of color in design. The need to reevaluate the relationship of color and design simply stems from a new dimension brought into play by chromogenics, namely the dimension of time. In other words, a dynamic range of colors instead of a fixed, static color is a relatively new condition for designers, worthwhile to reflect on, investigate, and discuss.

Although studies of perception date back to antiquity, in the recent past, a particular branch of anthropology, the anthropology of the senses, helped to clarify the mechanism that connects the sense organs to the mind and defined perception as the psychic-physical process that synthesizes sensitive data in forms with meaning. The anthropology of the senses has allowed us to understand that the

M. Ferrara and M. Bengisu, *Materials that Change Color*, PoliMI SpringerBriefs, DOI: 10.1007/978-3-319-00290-3_4, © The Author(s) 2014

sensory perceptions are subjective because they depend not only on physiological factors related to human sense organs but also on psychological factors relating to the interpretation of sensory stimuli in a process that links the brain to the sense organs. That is to say that cultural orientation allows space for the expression of individual sensitivity.

Perception depends not only on psychological factors and the individual's culture, but also on the basis of belonging to social groups and historical periods. Every society and every era has a different perception of colors so they are interpreted in a particular way in each case.

Humans experience things through their senses and according to their own culture that gives sensations their meanings, from which arise emotions, thoughts, and even empathy between what the designer thinks (his poetics) and the feeling experienced by the product's user. Meaning is associated with context, trends, culture, and psychology, even if today the globalization of visual culture and design is helping to determine a kind of objectivity in the interpretation of colors.

A designer may use red color on a certain product to get the attention of the customer or reader, but this may be associated with romance, eroticism, or passion as in European culture, or luck as in Chinese culture. Design studies have contributed a lot to the objectification of the meaning of color and on the perception of what designers are able to design, select and combine in terms of shapes and colors to achieve a certain expressive result, to attract attention, hide or highlight certain details. Even the media through which an international culture of color spreads help to form a social language and record changes in taste.

Although the meaning of color depends on many factors, its emotional and physiological effects can be determined scientifically. For example, warm colors like red, orange, and yellow are stimulating while cool colors such as blue and green are calming. In this connection we may recall the well-known experiment of Itten (1961). Itten empirically demonstrated that when stationed in two environments at the same temperature, but of a different color, the personal sensitivity to cold can change by 3–4 °C. In a red–orange room people started to feel cold only when the temperature falled down to 11–12 °C while they felt cold already at 15 °C in a blue-green room. This is explained by the effect of color on blood circulation; blue-green colors slow it down, while red–orange colors stimulate it. The experiment also demonstrated the difference between physical reality and phenomenal reality.

Warm colors are preferred by interior designers to draw attention during shopping or to give people a warm feeling in a ski resort when they enter the building from the freezing cold outside. However such colors are also known to make people feel somewhat tense (Zelansky and Fisher 1999; Ocvirk et al. 2006). Cool colors have a relaxing effect and that's why they are commonly used in healthcare facilities, rehabilitation centers, and correctional institutions. The practice of using textiles with colors of the national cuisine (from sausage to mustard and pickles) in German homes of the 1980s is also well-known.

These are just a few examples of how color is used by designers to evoke a certain feeling or sometimes even not to change one's emotional state at all. Colors are used by artists and designers to draw our attention (red traffic sign), to symbolize a certain value (the Valentine red as the object of desire; green packaging as environmental friendly), or to express a certain mood (a photograph with yellow and red autumn foliage to express melancholy). However this catalog of applied colors is continuously modified due to technical and cultural reasons.

Dynamic colors could change some of the conventional approaches to color. Sometimes we may feel bored to look at the same painting or photograph hanging in our living room for a long time. It would be interesting and stimulating to see its colors change slowly during the day. Followers of fashion like to change their clothes, shoes, and accessories with the introduction of a new season. Shoes that change color according to the time of the year could create a niche market. Keeping a harmony of colors would be a challenge for designers and artists in a photograph, painting, or product with dynamic colors.

Color changes also occur in nature and we are familiar with some of them, for example leaves turning yellow during fall. Such changes are slow and take a long time to complete. Fast color changes can be observed on the skin of various animals such as chameleons, frogs, and squids. This topic is discussed in Sect. 2.8.

Color movies, color TV, and computer graphics are three fields where dynamic colors play an important role in products. Some principles used for computer graphics related to issues such as contrast, perception, color appearance, and color design are surely useful when designing with chromogenic materials. However two aspects of computer graphics (as well as color movies and color TV) are significantly different from color in chromogenic materials. Firstly, such media work with rays of colored light which follow the principles of additive color. On the other hand, paints, pigments, dyes, or prints used for chromogenic materials obey principles of subtractive color (Ockvirk et al. 2006). Secondly, the speed of color changes are typically much faster in the three media mentioned above, compared to today's chromogenic media. From this perspective, it seems that designers could learn a lot from nature, especially from animals that can change colors quickly, for research, development, and design related to chromogenic materials.

4.2 Designing with Materials that Change Color

Today a diverse range of chromogenic materials are available in the market and an increasing number of designers are interested in them. Like all technological innovations, chromogenic materials are creative stimulants and they necessitate experimentation with regard to characteristics, technical behavior, and aesthetic possibilities. The experimentation can lead to small or large inventions, product innovations, new poetics of design, original insight to reality, and interesting interpretations of everyday life.

The design process is fuelled by a sense of curiosity and a desire to understand how far certain materials can be pushed and investigated. It is an abductive process which, as Charles Sanders Pierce (Eco et al. 1983) argued, is the only form of reasoning that improves our knowledge

Some projects celebrate the pure excess and physical abundance of the materials that change color, some the novelty (Fig. 4.1). Other schemes employ materials in surprising and unconventional contexts. At times, the concept stage of a project will be driven by the physical properties of materials. The behavior of these materials can often unlock problematic issues regarding the spatial arrangement or surface details and textures of a building.

The set of experiments carried out is an interesting scenario of *lateral thinking* that feeds the knowledge of technology and steers it towards a desirable innovation. The goal of this research scenario is to discover possible and desirable

Fig. 4.1 Greg Saul, *Thermochromic Lamp*, experimentation with thermochromic ink on paper. Courtesy of Greg Saul

relationships between the object and the end user. Experimentations deal with pointing out potential briefs and concepts that can address research design, problems, and feasible solutions.

The intrinsic dynamics of these materials, the capacity of continuous adaptation, and a harmonic transition make them extraordinary collectors of experience. In fact, by modification of color and transparency, the two most immediate among the visible aspect of objects, these materials offer new opportunities and the emotional involvement of the users. Many projects demonstrate the interest of designers for the concepts of dynamism or transition (Fig. 4.2) that the new materials carry within themselves.

The ability to interact with the environment in a more or less complex and adapting their visual appearance to environmental conditions makes these materials very attractive in various contexts of use, like interior design, architecture, and product design. Thanks to such potentials, they have also been employed in the field of art, such as in the work *Heart* by the Japanese artist Kiyoyuki Kikutake,

Fig. 4.2 Felice Limosani, *The Art of Transitions*, Transitions Optical production, 2012. This video-installation, exhibited in the space of QC Terme di Milano, explores the concept of *transition* to enhance the photochromic technology of Transitions® lenses. The installation does not apply photochromic lenses, but tells us the unconventional experience that they introduced in our lives

exhibited at the Tokyo Museum of Modern Art. The artwork, realized in steel and covered with thermochromic paint, allows the auto-regulation according to daily changes in temperature, changing the whole appearance from yellow to red. The work assumes one appearance in the morning and a different one in the evening, celebrating transitions of light, telling the story of changing time and seasonal cycles.

Textile design is one of the areas to which more energy was devoted to experimenting with smart materials in the last decade. This area of research sees the convergence between smart materials and electronic industry, representing the future of smart clothing, wearable computing, fashion and design. Even though the first research studies were lead by electronics development groups (in healthcare, entertainment, sportswear and communication), the concept of smart textile materials also had a significant impact on the design and fashion world. In smart textile research, the approach has gradually changed from a technical one to a user centered one. Moreover most researcher teams create the visions and product scenarios before developing any applications to ensure that intelligent functions meet the user requirements and the future lifestyle (Ariyatum and Holland 2003).

In the field of textile design, various studies have focused on the use of thermochromic textiles to develop visual communication devices, like the work of Maggie Orth, one of the first creative and technical practitioners of electronic textiles. Between 2003 and 2004, she has developed the project *Dynamic Double Weave* (Fig. 4.3), a color-change fabric, which makes use of a color mutation and repeated color changing patterns randomly generated (Orth 2013). She has explored the expressive potential of the fabric with the variation and repetition of colors through the use of control software.

With a similar intention Berzowska (2005), a textile designer who works in the field of electronic textiles, has designed *Shimmering Flower*, a visual display made of soft components, conductive yarn and fibers for power supply, communication and networking, material for display by the use of thermochromic pigment, Nitinol and electronic ink. The prototype represents a soft, washable, animated display. Actually this project is located on the border between design, art, haute couture and research, but new applications seem to arise from the development of smart systems.

Of great interest is also the research conducted by Linda Worbin at the Swedish School of Textiles, University of Borås, on *changing surface* and *dynamic patterns* (Fig. 4.4). As stated by the Textile Design Department of that university, the practical investigations are useful to become familiar with the basic aspects of those new *media*, their potential and practical limitations. Practical investigations serve to acquire fundamental skills, which include the ability to select, combine and transform colors, to apply the materials, and to see what is possible to do with them. The experiments, even if conducted at the level of crafts and do-it-yourself, are useful to develop a series of possible aesthetic expressions as examples from which to depart for new creative challenges (Hallnäs and Redström 2001). Their approach combines research in color-inspired design, in color technology and design/technology interface.

Fig. 4.3 Dynamic double weave

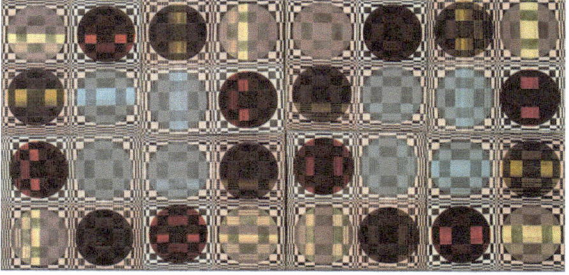

The studies commonly involve experimentation with electronic circuitry, to provide controlled and regulated stimulus, with the aim of producing dynamic and responsive artifacts. Overcoming this stage, the trials will be increasingly engaged in translations into rich interactions.

Industrial firms also start to become interested in chromogenic materials because they potentially permit design-driven innovation. For example Philips proposed the project Skin: tatoo (Fig. 4.5) within the program Design Probes dedicated in *far-future* research initiative to track trends and developments that may ultimately evolve into mainstream issues that have a significant impact on

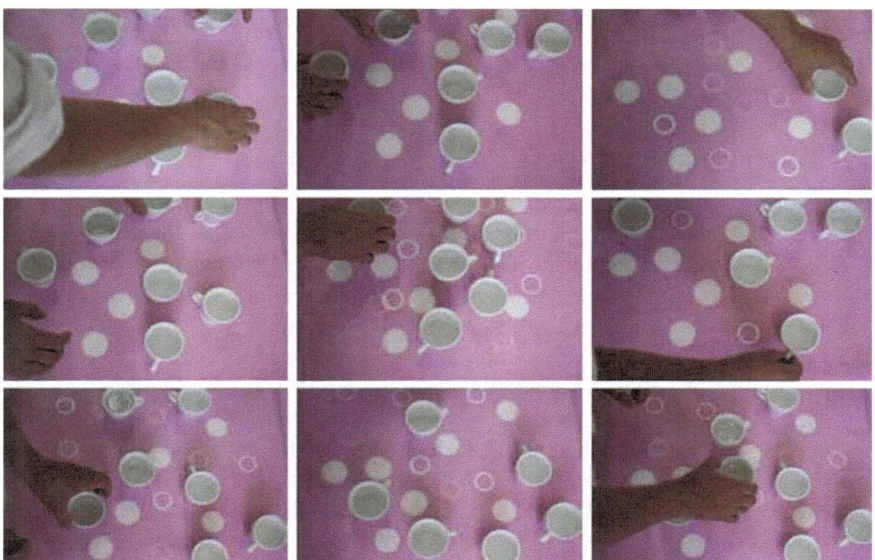

Fig. 4.4 Linda Worbin during her practical experimentation Disobedient Tablecloths with thermochromic textile at the Swedish School of Textiles, University of Borås. Courtesy of L. Worbin

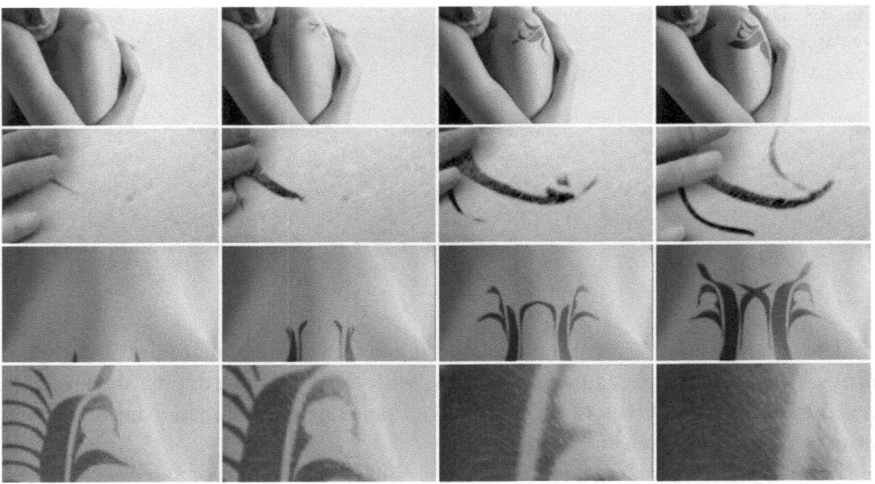

Fig. 4.5 Philips design, *Electronic Tattoo*, thermochromic ink, 2007. Courtesy of Philips Probe

business. Skin: tatoo is a thermochromic pattern which appears on the skin when touched, leaving a trace of a lover's embrace.

Studies so far undertaken demonstrate that these materials, if suitably applied, increase the functional performance, aesthetics, and communication skills of objects and media, while saving energy with respect to traditional systems (Chap. 5).

4.3 Theoretical and Methodological Considerations

Materials with dynamic characteristics present complex issues relevant for design. Their smart behavior presents new challenges to designers which involve the comprehension of functional, technical, and aesthetic values to be developed in the project using smart materials in order to answer questions such as which type of smart materials to use, as well as how and why to use them.

The composite nature of smart materials (support + ink or pigment) and the combination of electronic tools to manage the performance creates a series of interdependencies that makes the design process additionally complex (Worbin 2010). Therefore, applying materials that transform present theoretical and methodological problems that need to be addressed. The design practice challenge consists in managing the new potentials as changes of expression in our temporal dimension and interactivity paying attention to user experiences.

4.3.1 Design Primario, Design and Perception, Material Design

It can be maintained that chromogenic materials are potential instruments for *material design*. This term includes all those experiences from the activities of the Bauhaus, especially in the textile workshop, which have helped to define materials research, setting the question on the new possibilities of use and expressiveness of contemporary technology. A long tradition of this research has characterized Italian design, finding the peaks of particular interest in some of the practices and theories as *design primario (primary design)* (Branzi 1983, 1984), focusing on the *soft qualities* of products (including color, transparency, smell, sound, temperature, surface texture, their response to light and its variation, etc.).[1] The soft qualities, relating to material characteristics, their interaction with environmental conditions (intensity of the rays of natural light, etc.) are fundamental to the characterization of the objects and environments, indoors or outdoors. These are at the heart of human experience in the artificial environment because they affect the user's perception and thus the conscious or unconscious meaning transmitted by the design.

Starting in the mid-70s, design research based on these qualities, which in the age of *movimento moderno (modern movement)* were considered insignificant for production, enjoyed notable success so much so that they became the central theme of sophisticated industrial strategies of differentiation of a product in a saturated market. Shifting the focus from structural qualities (closely related to the mechanical properties of materials) and functional requirements, to soft ones, design research focused on the expressiveness of products, the physical-environmental- material experience, and the communicative and emotional value of materials with the aim of raising the quality of the relationship between man and the world around him.

This theory was a way out to overcome the crisis of plastic arts, in the years immediately preceding the birth of *design primario*, leading to the certainty that art could continue to exist even in the absence of compositional representation of shape.

Towards the end of the 1970 and 1980s, important research and projects on materials, color, graphical decoration, and textures were realized. These include studies on the color that merged into *Colordinamo*, a publication that addressed designers and manufacturers from various sectors, on an annual basis since 1975,

---

[1] *Design primario* is a design approach that was started in Milan at the Design Center Montefibre (Montedison group) between 1974 and 1977 by Andrea Branzi, Trino Clini Castles and Massimo Morozzi and involved other leading representatives of Italian design such as Mario Bellini, Alessandro Mendini, Denis Santachiara, and Ettore Sottsass.

which focussed on 40 selected colors in every issue since they are linked to the major cultural themes of the time, with technical and cultural information.[2]

Another interesting project is *Fisiolight*, a controlled-light curtain system by Montefibre Design Center, which was used to design the curtains of an internal space controlling the quality of light resulting from the interference between the colors of the curtain and the natural light that filters into the environment. The system was based on the decomposition of a chosen decoration in pixels of colors identified according to the *colorimeter*, an instrument specially designed for the control of color results.

The design of color characterized materials such as lacquers, polymers, upholstery fabrics, plastic laminates and many other coating materials that were replacing natural materials. In the same period a new norm was issued in Germany that abolished the use of monochromatic furniture systems in the workplace to improve the quality of these environments through the introduction of articulated color ranges. The aspects of color, texture, fonts and decorations, but also olfactory and acoustic aspects became elements of design. Starting from the idea of designing the identity of the new materials, a revolution was taking place of the languages that touched the relationship of use, the way of interacting with objects and places.

From this point also experience in *material design* was developed such as those based on *Meraklon Fibermatching System 25* that allowed to design needle-punched carpets made from synthetic fibers solution dyed in a theoretically infinite number of colors. The system gave rise to a rational method of definition of mixtures of colors, design of new colors and shades with a "recipe" of reference, using fibers of different color. The fibers of different colors appropriately mixed generated solid colors when viewed from a distance.

The concept of decoration was also recovered as a system of visual information, use and culture, rich of semantic characteristics. This concept was used as a base for the editorial project *Decorattivo*, decorative manuals.

Designer Trino Clini Castelli, who devoted himself to these issues, coined terms such as *emotional identity*, *aesthetic sustainability*, and *ecology of emotionality* to define the direction of research on surface design and give rise to a particular branch of design research called *qualistics*.

Following these Italian experiences, the sensory-perceptual approach to design has evolved thanks to the research of the Japanese designer Kenya Hara, currently creative director of Muji. He claims that "the field of design is human perception". During the preparation of the exhibition *RE-DESIGN: Daily products of the twenty first Century* (2007), which stirs many reflections on the functionality and aesthetics of everyday objects, Hara invited designers to redesign a given object. In his book, Designing Design, Hara defines his vision of the twenty first century as an age of discovery. While reminding us that the act of making the known

---

[2] The topics discussed were: the colors of energy (1975), pre-synthetic colors (1976), permanent colors (1977).

unknown is a creative act, Hara aims to nuance our understanding of the future by conceiving of it not as a radical break with the past but, rather, a rethinking of the ordinary.

The principle of Hara's method is to move the point of view of traditional design, focused on the shapes, sizes and functionalities of objects, towards the unconscious perception. About some of the projects of designer Naoto Fukasawa, Hara said: "the Fukasawa method is to examine our subconscious behaviour and design for that. This manner of thinking reminds me of an emerging cognitive theory called affordance. Affordance is the comprehensive understanding of both the subject of an action and the environment that affords, or allows for, a certain phenomenon. For instance, standing seems to be a behaviour imbued with the will of the subject (the standing person), but in reality, standing wouldn't occur if there were no gravity and no decently solid surface on which to do it ... In the case of standing, both gravity and solid surface are said to "afford" the action" (Hara 2007).

The concept of *affordance* was introduced by the American psychologist James Gibson who coined it in 1979 in his work *The ecological approach to visual perception* to define the physical quality of an object to prompt the user on the appropriate actions to manipulate and use it.

With the subsequent exhibition *HAPTIC. The awakenig of the Five Senses* (2007), Hara conducted a survey on the perception and capacity of design to stimulate *creative awakening of the human sensors.* "The concept of 'haptic,'" Hara described, "leads to the idea that we not only design form by creating a shape or an object, we also design how it feels. A human being is a bundle of delicate senses. Science doesn't only help the evolution of materials and media, it also helps us understand the senses, where there may be hidden a whole new, undiscovered territory ... 'Haptic' means another design attempt to expand the world atlas of senses." In occasion of this project, Hara poses to the designers invited to the project the aim of stimulating the sense of touch through the use of colors and shapes, before they even think about an appropriate form for the function to be performed. On this occasion, Hara said, "I believe that technology will bloom when planted on this sort of sensory perception."

The approaches mentioned in this section are the basis of the contemporary culture of material design that deserves to be reclaimed and reactualized by designers to tackle design with *changing materials*. The same is true for other experiences closer to artistic research, such as the *Kinetic and Programmed Art*, which employed real or unreal movement, and the variation of light and color according to a program prepared as a means of expression.

4.3.2 Designing the Interaction

Changing materials mark a turning point in the methods of design and material design. These new tools enhance the ability of sensory stimulation of design. The variability of their appearance offers new opportunities for research of perpection design and the poetic study of reality. But at the same time it causes an increase in the design complexity because the variables involved are multiplied; the temporal dimension is added to issues relating to shape, color and texture. It becomes insufficient to design the sensorial and perceptive aspects of materials in their fixity, but it is necessary to include the temporal dimension, considering the variability of sensorial aspects in time and the temporal form of interaction between product and user. This new aspect determines an intersection of product design and other trajectories of design research, such as *interaction design* and *experience design*.

Interaction design is a *user oriented design* method born in the field of computational design with the aim to design the digital interface. It examines the user-product interaction to facilitate the use of the computer based product and the interaction with the object. What characterizes the method is the design of the acts that in the time of use allow the appropriate use of objects and systems (Buchanan 2001) and the combination of the acts with computational technologies that become "material of design". The design process is defined as problem solving whose basic question is the definition of the problem and what it means to solve it (Coyne 2004; Friedman 2003; Schön 1983). Although this approach has its roots in computer science with the aim of solving design issues related to screens and digital interfaces, today the term has a much broader meaning. Interaction design aims to create a physical and emotional dialogue between a product and its user. This dialogue is expressed in the dynamic interplay between form, function, and technology (Kolko 2011). Hallnäs and Redström (2006) stated that interaction design can no longer be considered only as a subfield of computer science, but a link beetwen basic research in computer science and product applications for new expressive design materials. Incorporating a composite entity in both hardware and software, smart materials can no longer be regarded as neutral technical solutions but rich in expressive potential. The expressive result depends on a number of basic choices which have their profound aesthetic nature and of which the project is composed of.

Some results of this reflection on the design method can be seen in the practices of textile patterns that involve multidisciplinary teams, with different specific tasks (electronic engineers, textile designers, textile engineers, interaction designers, programmers, artists, and philosophers) who succeed to cover a broad spectrum of questions and involve new design variables. These include rates and methods in which the material is exposed to the stimulus (the environmental conditions that vary randomly and/or an intentionally applied input) and timing and mode of response to the stimulus; times and modes of interaction that occurs between the object and the environment and between the object and the human body. The design process is broken into pieces/blocks consisting of unique or multiple solutions to variables that are considered very basic components.

The experience design approach, which is an extension of *usability centered design*, is interested in the overall experience and the satisfaction of the user during the use of the product or system, with the aim of optimizing them. This approach considers the quality of experience of a specific user during the interaction with a specific product, and studies sensations, emotions, joy, simplicity, and ease of action during usage and also evolution of values and significance of use derived from experience, within a precise picture of social, material, and cultural context of reference, with a holistic vision (Battarbee 2007). Experience design encompasses the design of interaction through the involvement of all the senses and, to include the quality of interaction, also involves many disciplines such as perceptive and cognitive psychology, cognitive sciences (neuropsychology), linguistics, and semiotics. This complex methodological approach has been developed in the scope of software design, web applications, and digital devices. Today, experience design is one of the new focal points of research in product design.

One design example which declares to adopt experience design is the project of a smart vacuum cleaner of de Bruijn (2011) (Fig. 4.6). Describing the applied methodology during design, de Bruijn emphasizes the importance of designing the user experience, which anticipates the full involvement of senses to facilitate the understanding of the functioning of the object and the design of user-product interaction, based on the model of human–human interactions made of reciprocal communication and reactions to render the experience more satisfactory.

This and other examples of product design demonstrate that the application of chromogenic materials permit the improvement of communication with the product. Through the modification of color, induced by a stimulus and without the

Fig. 4.6 A.A. de Bruijn's concept applies materials that change color to indicate bag level on top of the vacuum cleaner and on the suction head to communicate that the vacuum cleaner is aspiring dust or the area is clean and the user can move to another location

need for screens, since the material itself acts as an interface, it is possible to convey messages and information to the users, for example communicating what is occurring inside the product or how to use it. Chromogenic materials open new windows of opportunity for augmenting the reality of interaction, making it more continuous, persistent, and coherent to feedback (Minuto et al. 2011).

During the last few years, studies in design research which experiment with the use of smart materials began to have profound implications on theories of design. In certain media of scientific-academic research, a new vision emerges which rejects the conventional design approach which considers the computational level independently from the operational level, and thus, from the physicality of objects. Experimenting with the possibilities of smart materials in electronic products, researchers like Dunne, Coelho and many others who work in the field of interaction and experience design, try to recompose the two aspects, namely computational and operational, designing two dimensions of materials in unison: the physical and the digital. Their objective is to give life to objects and space whose electronic operation becomes tangible and capable to generate rich aesthetic experiences, while they transmit digital information. As Dunne (2005) states, such an approach intends to retrieve the materiality of the object, reducing the gap between the analog and the digital world since materials used at the micro- or nanoscale started to be used to construct transistors inside sealed boxes, which gave these objects technical consistency, although they made their operation incomprehensible from that point on. The current approach could improve the affordance of the objects, recovering the material richness which was lost during the passage from atoms to pixels (Coelho et al. 2007).

The intelligent material is capable to communicate. It becomes the medium and physical representation of behavior and implies new product aesthetics. A further implication of this approach is the possibility that computer science and materials science converge in design, architecture, and fashion.

Because of all that, chromogenic materials will be the actors of a radical change, in the modes of design, in interacting with and perceiving objects and the media that surrounds them. Designers will make use of potentials of chromogenic materials, recovering the lack of fixity for the sake of communicative quality between product and user, taking into account the immediacy of perception, the simplicity of communication, and the compliance to forms of human interaction.

4.4   Technical and Economic Considerations

Materials that change color bring new life to conventional products and design practices. A common object such as a T-shirt takes new meaning with its thermochromic or photochromic graphic design; it becomes more interesting, more useful, more fun to wear. These materials will also lead to completely new products and services, which may be only ideas and concepts today or which have not been imagined by anyone yet. Manufacturers, product developers, and designers will certainly be interested in such new opportunities. However, without understanding some of the technical limitations and without the development of suitable specifications, there would be much room for frustration.

We have already discussed some of the technical considerations in Chap. 3 under Manufacturing and Processes. Here, additional information will be given related to the properties of some of the available chromogenic materials and technologies. Table 4.1 summarizes the important properties of various chromogenic materials.

4.4.1 Thermal Stability of Chromogenic Materials

An important property to consider in all types of applications is thermal stability. Even if the application does not involve external heating or cooling, the ambient environment may be sufficiently hot or cold to destroy an organic chromogenic molecule. Spirooxazines, which are photochromic substances, start to lose their stability at 200 °C (Luthern 2004). Similary, the thermal stability of thermochromic conjugated polymers (polyacetylenes, polydiacethylenes, polythiophenes, and polyanilines) is lost above 200 °C. Photochromic powder made from naphthopyranes or spironaphthoxazines can withstand temperatures from −40 to 250 °C (Ritter 2007). On the other hand, some thermochromic inorganic pigments are stable up to 500 °C (Seeboth 2008).

Designers should select the appropriate chromogenic material considering both the process temperature and the application temperature of the product in question.

4.4.2 Light Stability

Some chromogenic materials are inherently unstable and their chemical structure can degrade when exposed to UV light. This is especially the case for photochromic colorants. Fatigue resistance is a similar term to light stability. In photochromic materials, the term fatigue indicates the loss of photochromic response due to photodegradation (Baillet et al. 1995). The light stability and fatigue resistance of a photochromic compound are related to the total time of exposure to UV light, and to a lesser extent, to the number of photochromic cycling. In order to make these compounds less vulnerable, ink manufacturers develop stabilizers or additives. When properly stabilized, photochromic inks can be stored for years, but when printed on textiles or paper and exposed to sunlight, they last only a few months (Homola 2003). In addition to hindered light stabilizers, antioxidants, thermal stabilizers, and UV absorbers are used to increase light stability or photostability (Corns et al. 2009). Among photochromic compounds, spirooxazines possess the best photostability followed by spiro[1,8a]dihydroindolizines (DHIs), and spiropyrans (Luthern 2004).

Table 4.1 Properties of selected chromogenic materials

Type	Compound	Thermal stability	Light stability	Activation temperature	Switching time
Thermochromic	Inorganic pigments	≤ 500 °C	Good	70–127 °C (reversible) 120–500 °C (irreversible)	V_2O_5; coloration 1 s bleaching 6 s
	Microencapsulated leuco dyes	≤ 200 °C	Insufficient	0–130 °C	*
	Conjugated polymers	≤ 200°C	Insufficient	NA	*
	Liquid crystals	−30 to >100 °C	Very good	−30 to >100 °C	10 ms
Photochromic	Naphthopyrans	−40 to 250 °C	Good	–	≤ 1 min
	Spironaphthoxazines	−40 to 250 °C		–	NA
	Spirooxazines	*	Good	–	NA
	Spiro[1,8a]dihydroindo-lizines	*		–	NA
	Spiropyrans	*	Good	–	NA
Electrochromic	Liquid crystals	−30 to >100°C	Very good	–	<16 ms
	Conducting polymers	*	Very good	–	1–100 ms
	Inorganic materials	*	Very good	–	10–750 ms

Not available

Thermochromic compounds are also vulnerable to UV light. The light stability of conjugated polymers and leuco dyes are rather poor. Environmental factors such as the encapsulating matrix and the presence of oxygen and water significantly affect the light stability. Stabilizers have been developed for leuco dye-developer-solvent systems which almost completely prevented the light fading effect, although none of such systems displayed reversible thermochromism (Seeboth 2008).

Liquid crystals possess high photostability (Wen et al. 2005). They have been used safely for long periods in displays of watches, calculators, and TVs. Inorganic pigments also have good photostability but most of them are toxic (Seeboth 2008).

4.4.3 Activation Temperature

The term activation (or switching) temperature (TA) with respect to thermo-chromic materials refers to the temperature at which an adaptive transformation occurs. For example the activation temperature of liquid crystals can be adjusted to any temperature between -30 and 120 °C by suitable formulation (LCR Hallcrest 2012). TA of inorganic pigments can vary between 70 and 500 °C (Seeboth 2008).

Thermochromic leuco dyes are commercially grouped according to the temperature range of the selected application. There are three standard ranges: cold activated colorants are used on labels and packaging of refrigerated products. TA ranges from 10 to 15 °C where a transition from clear to color occurs at TA during cooling.

Body or touch activated colorants may be designed either to become clear and reveal the color image underneath or to change color triggered by the heat of the body or hand. Thus TA is fixed at 31 °C.

High temperature activated colorants change color at TA = 47 °C. This temperature is just below the pain threshold so the user is warned about a safety hazard (Chowdhury et al. 2012; LCD Hallcrest 2012).

A related issue is the adjustability of the activation temperature. Certain substances such as microencapsulated organic chromogenic pigments and liquid crystals can be formulated to exhibit the desired colors and TA. Many possibilities exist to precisely adjust TA and other properties of thermochromic liquid crystals. Some strategies include combining cholesteric and chiral nematics, components with right- and left-handed helical structures, and addition of nematics, racemates, and smectics with no liquid crystalline properties (Hallcrest 1991).

4.4.4 Switching Time

Switching time is an important parameter to consider in products and applications exploiting chromogenic materials. Switching time of inorganic and polymeric electrochromic materials lie in the range 10–750 ms and 10–120 ms, respectively

(Somani and Radhakrishnan 2002). Indolo[2,1-b][1, 3]benzoxazines have a switching time of ten- to few-hundred nanoseconds, which is one of the fastest switching time among photochromic compounds. They also have excellent photostability (Barkauskas et al. 2009).

Some applications specify very short switching times. For example, for LCD TVs, the switching time of liquid crystals must be below the frame time (the time span between each different frame) of 16 ms (Pauluth and Tarumi 2005). In some applications, relatively long switching times may be tolerated or even be preferred by designers. For example, the photochromic transition on a dress would be interesting to witness for someone watching it, hence the switching time of several seconds rather than milliseconds would be more appropriate.

Among inorganic materials, V_2O_5 electrochromic nanowires were shown to switch from red-brown to green in 1 s while bleaching (from green back to red-brown) took 6 s (Wang et al. 2010).

4.4.5 Economics

Due to highly specialized materials and processes used in the preparation of chromogenic materials, their price is significantly higher compared to comparable materials. For example, a gallon of UV-curable regular ink costs $100–200 while a thermochromic UV ink would cost about $800. Similarly, the price of a photochromic screen-printing ink is about four times the price of a regular screen-printing ink. The price for a 57 cm × 57 cm electrochromic window was $1,000 in 2009. The suitable price for market penetration is estimated to be $100–$250/m^2 (Pawlicka 2009). Polymer-based flexible smart glass technologies may reach this target and help the diffusion of architectural and other electrochromic applications.

Researchers, designers, and teachers may be interested in small quantities of chromogenic materials for experiments and projects. Temperature sensitive flexible vinyl sheets 15 cm × 15 cm in dimension and coated with liquid crystals are available for $20–$30 (Inventables 2013). Thermochromic and photochromic labels are sold for about $7 per sheet (30 labels per sheet—label size: 2.5 cm × 6.5 cm) (ThermometerSite 2013). Thermochromic powder pigments of different color which turn clear at 30 °C are available at a price of ∼$30 for 20 g boxes (Amazon 2013). These affordable prices make chromogenic materials more interesting for new designs and applications.

References

Amazon (2013) Thermochromic pigment: changes color with temperature. http://www.amazon.com/Thermochromic-Pigment-Changes-Temperature-Yellow. Accessed 15 July 2013

Ariyatum B, Holland R (2003) A strategic approach to new product development in smart clothing. In: Proceedings of the 6th Asian design conference, Tsukuba, 2003

Baillet G, Giusti G, Guglielmetti R (1995) Study of the fatigue process and the yellowing of polymeric films containing spirooxazine photochromic compounds. Bull Chem Soc Jpn 68:1220–1225

Barkauskas M et al (2009) Ultrafast dynamics of photochromic compound based on oxazine ring opening. Lith J Phys 48:231–242

Battarbee K (2007) Co-experience: Product experience as social interaction. In: Schifferstein HNJ, Hekkert P (eds) Product experience, Elsevier, Amsterdam, pp 461–476

Berzowska J (2005) Electronic textiles: wearable computers, reactive fashion, and soft computation, textile, vol 3(1). Berg, UK, pp 2–19

Branzi A (1983) Il design primario. In: Branzi A Merce e metropoli, EPOS, Palermo, p 71–76

Branzi A (1984) La casa calda. Idea Books Edizioni, Milano

Buchanan R (2001) Design research and the new learning. Des Issues 17(4):3–23

Chowdhury MA, Butola BS, Joshi M (2012) Application of thermochromic colorants on textiles: temperature dependence of colorimetric properties. Color Technol 129:232–237

Coelho M, Sadi S, Maes P, Berzowska J, Oxman N (2007) Transitive materials. Towards an integrated approach to material technology. In: Workshop Proceedings of ubicomp: international conference on ubiquitous computing. September 2007, Innsbruck, Austria, pp 495–500

Corns SN, Partington SM, Towns AD (2009) Industrial organic photochromic dyes. Color Technol 125:249–261

Coyne R (2004) Wicked problems revisited. Des Stud 26(1):5–17

de Bruijn AA (2011) Enriched expression of smart materials in consumer products. Design of a smart vacuum cleaner, Master Thesis, Faculty: Engineering Technology, Master programme: Industrial Design Engineering, Department: Design

Dunne A (2005) Electronic products, aesthetic experience and critical design. MIT Press, Cambridge 5

Eco U, Sebeok TA (eds) (1983), Il Segno dei Tre: Holmes, Dupin, Peirce, Bompiani Milano

Friedman K (2003) Theory construction in design research: criteria: approaches, and methods. Des Stud 24(6):507–522

Gibson JJ (1979) The ecological approach to visual perception. Houghton Mifflin, Boston

Hallcrest (1991) Handbook of thermochromic liquid crystal technology. http://www.hallcrest.com/downloads/RT006%20randtk_TLC_Handbook.pdf. Accessed 11 Oct 2012

LCR Hallcrest (2012) Thermochromic technology. http://www.colorchange.com/thermochromic. Accessed 27 Sept 2012

Hallnäs L, Redström J (2001) Slow technology. designing for reflection. Pers Ubiquit Comput 5:201–212

Hallnäs L, Redström J (2006) Interaction design. Foundations, experiments. Textile research center, The swedish school of textiles, University College of Borås and Interactive Institute, Sweden

Hara K (2007) Designing design. Lars Müller Publishers, Zürich

Homola J (2003) Color changing inks. http://www.xslabs.net/color-change/how-stuff-works.htm. Accessed 22 Oct 2012

Inventibles (2013) Temperature sensitive flexible sheets. https://www.inventables.com/technologies/temperature-sensitive-flexible-sheets. Accessed 15 Jul 2013

Itten J (1961) The art of color: the subjective experience and objective rationale of color. Van Nostrand Reinhold, New York

Kolko J (2011) Thoughts on interaction design. Morgan Kaufman, MA

Luthern JL (2004) Photochromic and thermochromic colorants. In: Charvat RA (ed) Coloring of plastics, vol. 1: fundamentals. Wiley, NJ

Minuto A, Vyas D, Poelman W, Nijholt A (2011) Smart material interfaces—a vision. In: Proceedings of 4th international ICST conference on intelligent technologies for interactive entertainment. (INTETAIN '11) LNCS Springer, Genoa, Italy

Ocvirk GO, Stinton RE, Wigg PR, Bone RO, Cayton DL (2006) Art fundamentals. McGraw-Hill, NY

Orth M (2013) Maggie Orth. http://www.maggieorth.com. Accessed 26 July 2013

Pauluth D, Tarumi K (2005) Optimization of liquid crystals for television. J SID 13:693–702

Pawlicka A (2009) Development of electrochromic devices. Recent Pat Nanotechnol 3:177–181

Ritter A (2007) Smart materials in architecture, interior architecture, and design. Birkhäuser, Basel

Schön DA (1983) The reflective practitioner: how professionals think in action. Temple Smith, London

Seeboth A, Lötzsch D (2008) Thermochromic phenomena in polymers. Smithers Rapra, Shropshire

Somani PR, Radhakrishnan S (2002) Electrochromic materials and devices: present and future. Mater Chem Phys 77:117–133

ThermometerSite (2013) Thermosmart for teachers. http://www.thermometersite.com/thermographics-for-teachers/view-all-products.html. Accessed 15 July 2013

Wang J, Sun XW, Jiao Z (2010) Application of nanostructures in electrochromic materials and devices: recent progress. Materials 3:5029–5053

Wen CH, Gauza S, Wu ST (2005) Photostability of liquid crystals and alignment layers. J Soc Info Display 13:805–811

Worbin L (2010) Designing dynamic textile patterns. Dissertation, Chalmers University of Technology, Sweden

Zelansky P, Fisher MP (1999) Colour. Herbert Press, London

Chapter 5
Case Studies

Abstract This chapter presents a number of case studies: products, projects, concepts, experiments, and smart systems using chromogenic materials. These were chosen based on their capacity to represent state of the art of the experimentation and application of materials that change color in different fields including product, interior, fashion, packaging, and visual communication design. They are useful to understand both the functional and expressive nature of these materials, by showing design methods and their results. They show the new qualitative dimensions that smart materials bring into industrial and product design, the role that these new materials and technologies can play, and their influence in different areas of product design. This general scenario of applying chromogenic materials conducted by designers in different parts of the world until today will help to answer questions like who, why, and how about design with *materials that change color*.

Keywords Photochromic · Thermochromic · Electrochromic · Hydrochromic · Biochromic · Smart systems · Interaction design · Product design · Interior design · Furniture design · Fashion design · Packaging design

5.1 Designing with Photochromic Materials

5.1.1 Sun-Reactive/Water-Reactive Dresses and Swimsuits by Amy Winters

Amy Winters is a young designer who is pushing the limits of fashion design by the use of smart textiles and interactive clothing for the entertainment, fashion, and advertising industries. Her brand, *Rainbow Winters*, showcases several collections of innovative dresses, skirts, and swimsuits, which change color under sunlight, water, or both. Winters has been developing smart textile applications for visual arts, performing arts, and the fashion industry since 2007. Specifically, these designs are intended for music videos, rock concerts, award ceremonies, special events, and advertisement (Rainbow Winters 2012).

New dimensions are added to garments and swimsuits through the use of hydrochromic and photochromic inks. Most of the sun-reactive applications in these collections involve screen-printed photochromic ink which transforms from clear to purple under daylight. Some examples can be seen in Figs. 5.1, 5.2, 5.3, 5.4 and 5.5. The first example shows an orange stretch-cotton mini-dress with petal sleeves. Photochromic ink, manually screen-printed on the dress, changes from clear to purple in sunlight. The transformation is clearly visible by the contrasting purple features on the orange background (Fig. 5.1).

A similar effect was used on a skirt with triangular panels, displaying lightning graphics created by a photochromic dot pattern (Fig. 5.2).

The orange nylon taffeta panels were prepared by dye sublimation printing. As seen from the figure, parts of the triangles changed from light colors to a darker purple when exposed to sunlight. The same effect was used on several swimsuits, one of which is shown in Fig. 5.3. This blue polyester/elastane swimsuit was color-gradient sublimation printed with sun-reactive dot patterns. Parts of the fabric change from clear to purple in sunlight.

Winters also used hydrochromic inks in some of her designs. These inks change color when they come into contact with water. The *rainforest* showpiece uses both photochromic and hydrochromic effects (Fig. 5.4). The photochromic print on the flowers transforms from clear to purple in sunlight. A different approach was used to demonstrate the hydrochromic effect on this dress. Some of the flowers and the bodice of the dress were treated with a white hydrochromic ink overlay.

Upon contact with water, this layer becomes transparent and reveals the color of the print underneath. This idea was also used on a swimsuit (Fig. 5.5). When the *rainforest* swimsuit gets wet, white streaks disappear and the whole swimsuit becomes purple. Although hydrochromic and photochromic effects are rarely used in fashion design nowadays, many more applications are bound to follow in the near future with the help of marketing, diffusion of knowledge, and cost reduction.

5.1.2 Drivewear Photochromic Lens by Younger Optics

Self-darkening sunglasses are probably the best known type of photochromic products since they have a relatively early market exposure, and since they are consumer products rather than scientific or technical products. Such sunglasses have been around in the market since 1966, the date when Corning first introduced the Photogray® lens (Van Gemert 2000). These lenses were made of a borosilicate glass containing fine crystals of copper doped silver halide. Under UV-light, copper (1+) ions give up an electron, which are used by silver cations to form

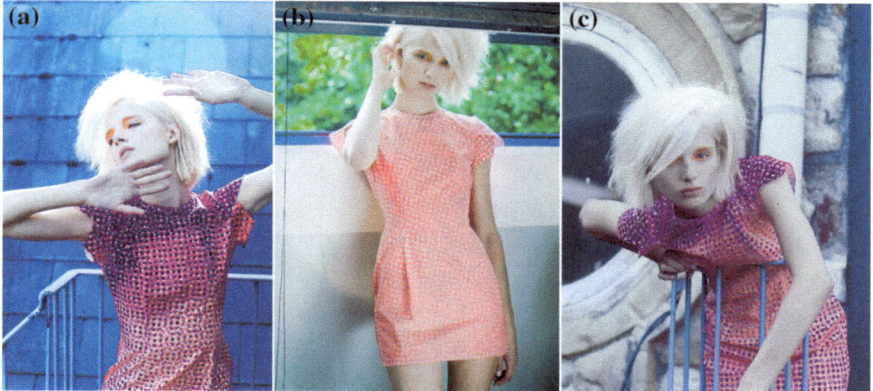

Fig. 5.1 *Orange* Petal Dress; stretch-cotton hand-screen printed. Photochromic screen-print changes from *clear* to *purple* in sunlight. Metallic gold *back*-zip and *pink* satin trim. Photo by Cereinyn Ord. *Courtesy* Amy Winters

Fig. 5.2 Tropical Storm Skirt; *orange nylon* taffeta, triangular panel, sublimation print. Photochromic screen-printed design changes from *clear* to *purple* in sunlight, here shown after exposure. Photo by Cereinyn Ord. *Courtesy* Amy Winters

elemental silver. Silver atoms combine as visible clusters which causes the darkening. The process is reversible, so when the UV-light is reduced or eliminated, silver atoms give back one electron and copper (2+) ions are formed. Such lenses still exist today but plastic photochromic lenses reign the market since the 1990s (Erickson 2009). The use of glass lenses for eyewear became a tough challenge after the introduction of new standards by FDA of USA to prevent eye injuries in 1972 because these standards mandated much thicker, thus heavier, glass lenses than used at the time. This meant uncomfortable glasses with thick and

Fig. 5.3 *Blue* Lightning Swimsuit; *blue* polyester/elastane colour-gradient sublimation printed. Triangular paneled lightning print, central panel printed with sun-reactive dot pattern. Photochromic ink screen-print changes from *clear* to *purple* in sunlight, here shown after exposure. Photo by Cereinyn Ord. *Courtesy* Amy Winters

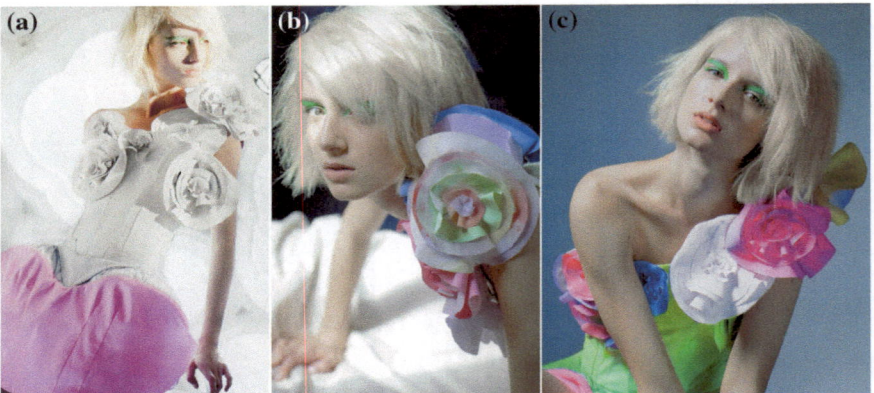

Fig. 5.4 Rainforest Dress; sculptural tulip-shape dress with 3-D flowers. Flowers multi-colored, *pink* and *blue* with *white* hydrochromic ink overlay and photochromic ink. *Green* color-gradient print bodice with *white* hydrochromic ink overlay, machine embroidered panels and *pink* skirt. Hydrochromic ink becomes transparent on reaction to water revealing the colored print underneath. Photochromic flowers change from *clear* to *purple* in sunlight. Photo by Cereinyn Ord. *Courtesy* Amy Winters

Fig. 5.5 Rainforest
Swimsuit; *purple* polyester/
elastane color- gradient
sublimation print. *White*
graphic screen print in
hydrochromic ink. The *white*
print disappears on reaction
to water, here shown before
exposure to water. Photo by
Cereinyn Ord. *Courtesy* Amy
Winters

heavy lenses. This challenge was met by the introduction of plastic lenses by PPG Industries, which employed a thermoset with the trademark CR-39®. Plastic photochromic lenses were introduced to the market in 1982 by American Optical with the trademark Photolite®. In these lenses, indolinonaphthoxazines were used for the darkening effect. This product was not successful in the market due to the blue color produced in the darkened state, instead of the more neutral gray or brown colors people were accustomed to. Subsequent generations employed naphthopyrans and indenonaphthopyrans. The molecular structure of naphthopyrans change under UV-light by the breakage of a weak chemical bond. The rearranged molecules absorb visible light. While these sunglasses perform well in daylight, they cannot provide the same effect inside the car because windshields contain UV-blocking agents. This was the driving force behind the development of Drivewear® by Younger Optics in collaboration with Transitions Optical.

Driving is a complex activity involving multitasking and the processing of a great deal of information in a short time. This situation is further complicated by changes in road conditions, weather condition, traffic conditions, and occasional obstacles (Fig. 5.6). All such information is captured by the driver's eyes. Therefore, unfavorable light conditions such as bright light or glare create risks for the driver. The combination of Transition Optical's photochromic technologies and Younger Optics' polarizing solutions provide optimum vision when driving in low light or full sunlight, or when the user is outside the car (Younger Optics 2012).

Advanced photochromic dyes provide auto-darkening effect both inside and outside the car, responding to visible as well as UV radiation. The polarization

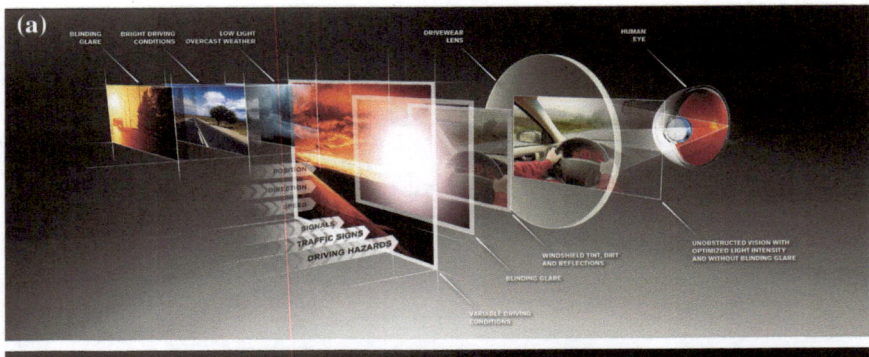

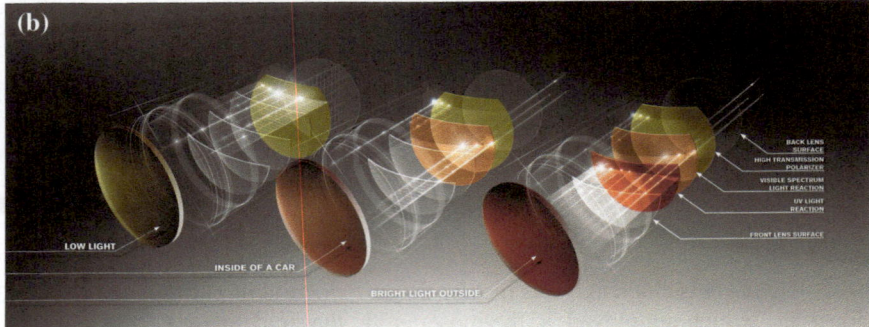

Fig. 5.6 Challenging driving conditions and their perception by the eye (*top*). The state of different layers of Drivewear® lenses under various light conditions (*bottom*). *Courtesy* Younger Mfg. Co

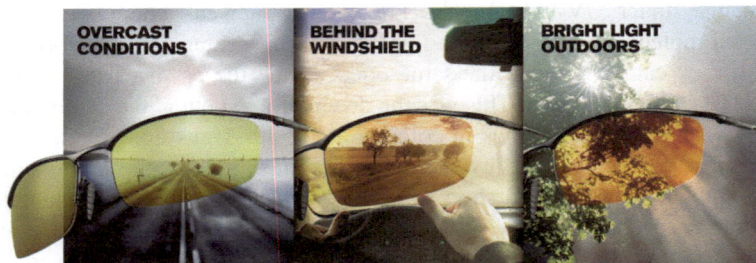

Fig. 5.7 Simulation of Drivewear® lenses under different outdoors and driving conditions. *Courtesy* Younger Mfg. Co

technology adapted to Drivewear® has the ability to work well not only when the lenses are dark, as in competing products, but also when there is overcast weather and the lenses are clear (Figs. 5.6, 5.7). Under overcast low light conditions, Drivewear® lenses allow high light transmission and prevent any possible glare.

The color at this stage is green to yellow. When bright sunlight hits the windshield, a smaller fraction of the light is allowed to pass through the lenses, now copper-colored, relieving the eyes from strain. Outside the car, the lenses become dark reddish brown and achieve their darkest color. Glare is prevented with polarized lenses under all conditions.

5.2 Designing with Thermochromic Materials

5.2.1 Thermochromic Brush by Corioliss

How can a simple product such as a hairdresser's brush be converted into something new and more functional? The answer is embodied in the thermochromic brush, introduced by Corioliss. Those who have experience with hair tongs and curlers know well that when the hair styling tool is hot, it is much more efficient at styling. It is possible to control the temperature in electrical tools such as tongs and curlers through microprocessors and thermostats. The temperature of an electrical device can be displayed with a digital indicator. However this approach is not possible in a regular brush since it is not supplied with electricity. A thermochromic coating is a good solution without the need for an electrical input. The brush is made up of a light ceramic barrel covered with a thermochromic coating which turns ivory with the heat of the hot air blown by a hair dryer. This color indicates that the brush is at the optimum temperature (about 50 °C) for styling. When the brush is cooled down to room temperature, its color progressively turns to black in a few seconds (Fig. 5.8). A change from ivory towards black may

Fig. 5.8 Thermochromic brush when hot (*left*) and at room temperature (*right*). *Courtesy* Corioliss, www.corioliss.com

Fig. 5.9 Partially heated
barrel. *Courtesy* Corioliss

indicate that the hair dryer is not pointing at the brush correctly or that the brush is
only partially heated, which may be a helpful hint for the user (Fig. 5.9).

5.2.2 In Heat, Installation by Jürgen Mayer

German architect Jürgen Mayer is one of the outstanding personalities in con-
temporary design thanks to his unique architectural achievements. His firm J.
Mayer H. Architekten realized many unique projects including the refectory at the
Karlsruhe University of Applied Sciences (2007), ADA 1—Office Building at
Alster, Hamburg (2007), rest stops in Gori, Georgia (2011), café and sculpture in

Fig. 5.10 General plan of
the installation. *Dark* areas
indicate thermosensitive
surfaces (Photographer:
Mauro Restiffe. *Courtesy* J.
Mayer H. archive, Berlin)

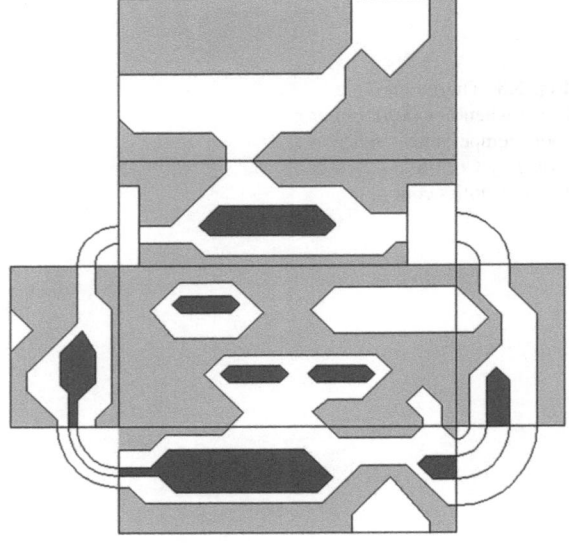

St. Barbara's Park Batumi, Georgia (2010), Metropol Parasol—redevelopment of Plaza de la Encarnacion, Seville, Spain (2011), and Court of Justice—Hasselt, Belgium (2011). However, he is not only involved in architecture. His work spans a wide range of fields from art to philosophy to design. Two of his interesting preoccupations are data security patterns and thermosensitive surfaces (Abrahamson 2009). These two themes appear in many of his installations and objects. Mayer describes data security patterns as "multi-leaf forms that courier services use" for data monitoring. In an interview, he explained that he used these patterns together with heat-sensitive print in a guestbook, in one of his exhibitions. Thus, the patterns became visible when heated, for example by the heat of the visitor's hand that wrote a comment, and disappeared once the visitor's hand moved away from the guestbook. To him, these patterns, which he collects, represent "strategic ornamentation" symbolizing volatility or evanescence of things (Voermanek 2012).

Thermosensitive surfaces appear in most of his installations as metaphors just like data security patterns. One example is *In Heat*. This is an installation which was exhibited at the Henry Urbach Architecture (HUA) Gallery, New York, in 2005. Certain parts of the walls and seats in this installation incorporated

Fig. 5.11 In Heat, general view of the installation. HUA Gallery, New York, 2005. Photo by Mauro Restiffe, *courtesy* J. Mayer H. archive

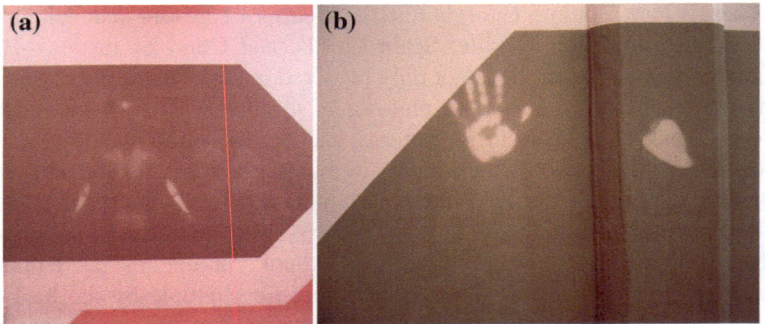

Fig. 5.12 In Heat, details from walls with thermosensitive paint. Photo by Mauro Restiffe, *courtesy* J. Mayer H. archive

thermosensitive paint, which reacts to touch or body heat by losing color (Figs. 5.10, 5.11, 5.12).

The thermosensitive surfaces represent a three-dimensional painting where "the viewer, creating a temperature shadow by touching, melts into the overall exhibition design" (Fig. 5.11). *In Heat* was inspired by Friedrich Kiesler's exhibition in 1947 presented at the Hugo Gallery in New York. Kiesler, who was also an architect, aimed to remove the boundaries of the wall, the floor, the ceiling, and merge them into a seamless, endless surface and space. He even included the viewer and made her part of the installation. Mayer's work took Kiesler's ideas one step further. With the help of new materials and techniques, he was able to really create seamless internal and external surfaces. Furthermore, with the help of thermosensitive surfaces, a fourth dimension can be experienced: time. The visitor, just like it was in Kiesler's Blood Flames exhibition, becomes part of the

Fig. 5.13 LIE,
Thermosensitive bed linens.
Photo by Christina
Dimitriadis, *courtesy* J.
Mayer H. archive

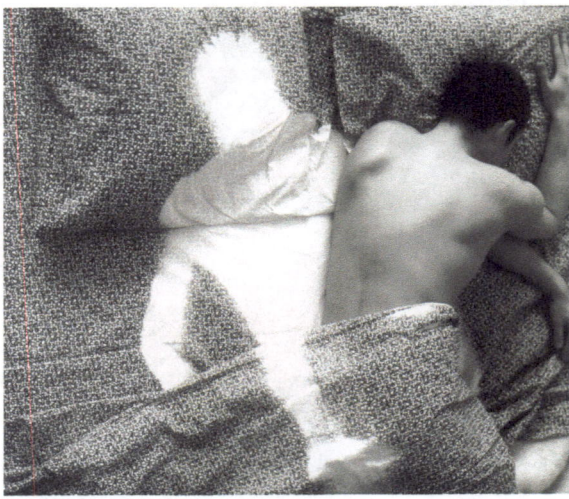

installation by leaving an imprint, a "temperature shadow" on thermosensitive surfaces. Thus, the viewer "melts into the overall exhibition design" (Mayer 2005; Cory 2007).

5.2.3 LIE, Thermosensitive Bed Linens by Jürgen Mayer

Some of Mayer's works involve a smaller scale, the scale of everyday objects. Most of these objects incorporate new materials, technologies, and ideas. These include sunglasses, elastic glass mosaic furniture, glass mosaic with digital data patterns, carpets with data security pattern, temperature sensitive furniture, and temperature sensitive bed linens. The latter is a limited edition set of bed linens which incorporated thermosensitive cotton textile (Fig. 5.13). Here again, one can notice both the data protection patterns and the thermosensitive effect combined in one product, once more symbolizing the transient nature of our daily activities, motions, and emotions.

5.3 Designing with Electrochromic Materials

5.3.1 Electrochromic Smart Glass by SmartGlass International

Smart glass is a general term which can have different meanings depending on the market and related technology. A practical definition could be: "adaptive glass which incorporates smart materials designed for transparency- and color-control". A large market exists for smart glass applications in exterior and interior fenestration and transportation industries because they offer significant benefits with regard to sustainability, energy consumption, comfort, and innovation. An important application of smart glass is in the control of light passing through a window and the interior temperature levels. Smart glass technology is an efficient and elegant solution for solar control. Various options exist for such applications.

Thermochromic windows allow more ambient light to the interiors by automatically becoming more transparent when it gets colder in a building or they block the incoming radiation when the internal temperature is above optimum. Photochromic windows help to optimize the internal temperature by becoming darker when the sun is bright or lighter when there is an overcast sky, just like photochromic sunglasses. A third option is electrochromic glass technology. According to a study undertaken by Madison Gas and Electric Company (2012), electrochromic windows provide the highest energy efficiency among the three chromogenic technologies, because they require the lowest energy for lighting and cooling. Electrochromic windows also offer the advantages of diminished glare and lower reflections in computer monitors (Granqvist 2010).

Fig. 5.14 Subdivision inside the Microsoft Head Office, Lisbon in the switched off (*left*) and switched on modes (*right*). *Courtesy* SmartGlass International

SmartGlass International has been active in the smart glass business since 2003 (SmartGlass International 2011). The firm offers two electrochromic smart glass solutions: LC SmartGlass™ and SPD SmartGlass™. LC SmartGlass is a privacy glass developed for interiors. In the "on" position, liquid crystal molecules align and allow natural or artificial light to pass through, making the glass transparent. In the "off" position, the molecules are randomly oriented, whereby most of the light is blocked and privacy is attained. Examples of application include meeting rooms in offices, shower cabins in hotels, teller and cash counting screens in banks, and private partitions in restaurants.

The Microsoft Head Office in Lisbon employed LC SmartGlass to create subdivisions of office spaces designed for meetings and presentations (Fig. 5.14). The use of smart glass panels eliminated the need for traditional curtains or blinds which can be difficult to clean and costly to maintain.

The use of smart glass also brings important advantages to hospitals. Many hospitals use curtains to separate patients from other patients or from visitors. Studies have shown that a significant fraction of such curtains contains bacteria such as *Clostridium difficile* (CDIFF), Methicillin-resistant *Staphylococcus aureus*

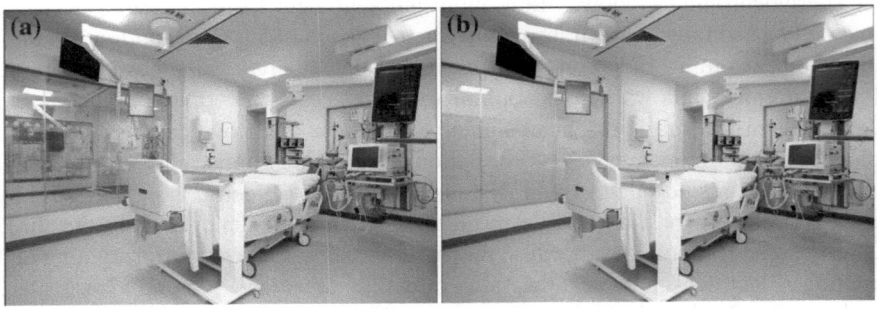

Fig. 5.15 Partitioning system used at Royal Sunderland Hospital in the switched on (*left*) and switched off modes (*right*). *Courtesy* SmartGlass International

Fig. 5.16 Interior view of the Globe of Science and Innovation, Geneva. Its electrochromic glass structure in the switched off/*dark* (*left*) and switched on/*bright* (*right*) modes. *Courtesy* SmartGlass International

(MRSA) and Vancomycin-resistant enterococci (VRE) (SmartGlass International 2011). Such bacteria can contaminate the hands of health care workers or visitors who touch these curtains, causing the germs to spread. The Royal Sunderland Hospital in Sunderland, UK employed LC SmartGlass in their Integrated Critical Care Unit to eliminate the possibility of contamination between two sections located in this unit: the operating room and the nursing room. Toughened smart glass door panels allow for patient privacy with the help of a switch, also permitting care staff to effectively control patient activity (Fig. 5.15).

SPD SmartGlass offers solar control for buildings. The production process involves lamination of suspended particle films between two or more glass panels 5–6 mm in thickness. Upon activation of the smart glass with a small current, rod shaped suspended particles become aligned, allowing light to pass through the panel. When the power is switched off, suspended particles return to their random alignment, which block 99.4 % of the light. This way it is possible to control glare, solar heat gain and UV exposure.

Various architectural projects already employed SPD SmartGlass technology. An interesting example is the Globe of Science and Innovation at Geneva, Switzerland. The Globe was originally built on the banks of Lake Neuchâtel for Expo 2002. The Swiss Confederation donated it to CERN as a venue for scientific, cultural, and educational activities (Foundation for the Globe of Science and Innovation 2012). The wooden globe is a symbol for sustainable development and serves as a landmark for CERN (Fig. 5.16). The top of the dome hosts a glass structure for bright natural lighting. For certain activities such as film shows or slide presentations, a dark environment is required. CERN wanted to keep the atmosphere and the architectural beauty of the Globe undamaged during the application for solar control. The SPD SmartGlass technology proved to be a suitable solution for this challenging application. When a presentation is taking place, even during the day, almost complete darkness can be achieved without the need for window coverings or other cumbersome means. Manual or automatic

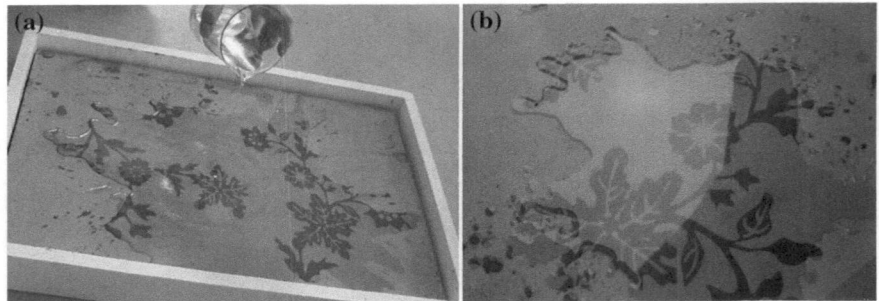

Fig. 5.17 Solid Poetry concrete tiles reveal a hidden pattern when they come in contact with water

tuning is possible to control the amount of light, glare, and heat passing through the window.

5.4 Designing with Hydrochromic Materials

5.4.1 Solid Poetry

Solid Poetry is the concept developed by designers Susanne Happle and Frederik Molenschot at droog® lab, in collaboration with the Dutch firm Terratorium, producer of ceramic and natural stone tiles. The prototypes of the project were presented at Milan's Saloni del Mobile in 2006.

The design research of the two designers according to the needs of the partner firm resulted in the possibility of using a hydrochromic ink in combination with concrete. From this idea, designs for floors and other building components with a poetic magic were born. The dull, gray, and neutral surfaces of these components made of concrete, unexpectedly reveal textures, scripts, or images when their surfaces get wet or humid.

The hydrochromic ink sprayed on the surface of concrete hides the underlying graphic, earlier applied on the concrete by screen printing. As soon as the surface becomes wet, the hydrochromic ink becomes transparent and it gradually lets the decoration below to appear. Once the surface is dry, the graphic will be invisible once again (Fig. 5.17).

Applications of this concept concern construction elements such as tiles in various forms, for floors and coverings suitable for external use, such as gardens and public spaces, where exposure to rain could offer viewers unexpected experiences, thanks to the dynamic qualities of concrete surfaces. Sanitary spaces for internal environments exposed to moisture, such as bathrooms and kitchens may be designed with patterns according to customer requirements.

The product offers a narrative quality, almost sentimental, with the capacity of enriching daily experiences.

5.5   Designing with Biochromic Materials

5.5.1   TopCryo™ Time–Temperature Integrators by Cryolog

Smart packaging is a rapidly growing field thanks to advances in smart materials, microelectronics, optoelectronics, and other technologies. Smart packaging solutions aim to facilitate traceability, record keeping, and sustainability in the food supply chain (Potter et al. 2008). Smart packaging is not limited to food packaging; it involves many fields other than the food industry. For example in medicine, smart indicators are very useful for the monitoring of blood, vaccines, medical implants, and drugs during storage and transfer. Some industrial products such as paint and sensitive chemicals must be protected from freezing; otherwise they will be damaged and unsuitable for use. Many smart packaging applications rely on visual indicators. Such indicators are a fast and direct way to inform the producer, distributor, retailer, or consumer about the condition of the product. Most labels or indicators use a color change to warn the customer when the product is not safe to consume anymore. Several types of devices have been developed so far including chemical, temperature, biological, and microbial indicators. The choice depends on the specific application.

Time–temperature indicators (TTIs) are valuable tools to track the history of a food product especially if there is a cold chain involved. They keep track of the temperature history of the product and indicate if a temperature abuse has occurred

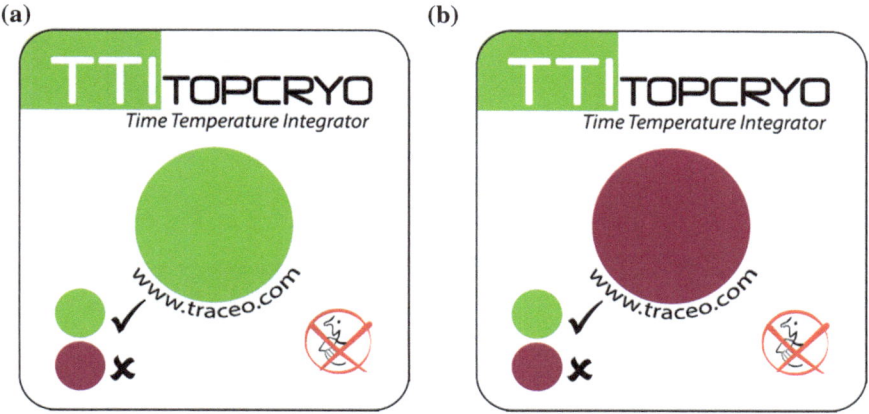

Fig. 5.18 TopCryo™ time–temperature integrator showing food product suitable (*left*) and unsuitable (*right*) for consumption (Courtesy of Traceo-Cryolog)

at any point in the cold chain (Oriakhi 2009). TopCryo^{TM}, developed by Cryolog, a company based in Nantes, France, is a time–temperature integrator, designed as a simple label to interpret the impact of time and temperature on heat- and time-sensitive products during their trip in the cold chain. An irreversible and easily measurable color change from green to red occurs on the label to indicate a discontinuity or abuse in the cold chain (Fig. 5.18).

The main benefit of TopCryo^{TM} and similar indicators is the protection and satisfaction of the final customer. However, there are other important benefits which interest the producers, distributors, and retailers. Improved product quality, reduction of waste, optimization of cold chain logistics, and increased profit from good cold chain management are some of these benefits.

TopCryo^{TM} is a microbiological indicator which simulates the spoilage of a food product. These types of indicators contain microorganisms within the label in order to measure the cumulative exposure to time and temperature. The growth of microorganisms in the label result in the color change from green to red in order to indicate that a similar process occurred in the product within the package (Potter

Fig. 5.19 Application of TopCryo^{TM} labels on packaged shrimp (Courtesy of Traceo-Cryolog)

Fig. 5.20 TopCryo^{TM} label is *green*, indicating that the meat is suitable for consumption (Courtesy of Traceo-Cryolog)

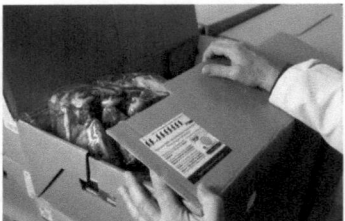

Fig. 5.21 TopCryo™ label is *red*, indicating that the clams are spoiled and *not* suitable for consumption (Courtesy of Traceo-Cryolog)

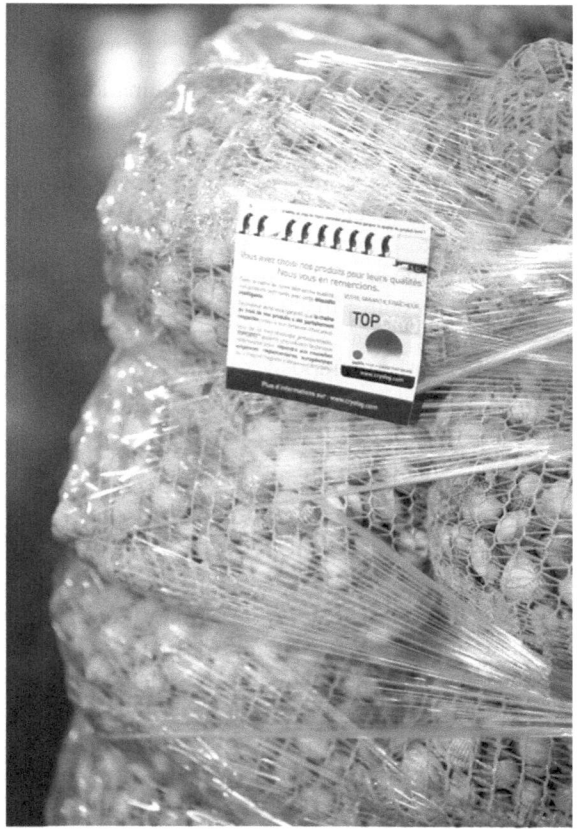

et al. 2008). The indicator is calibrated biologically according to the user's time–temperature specifications. Earlier versions of microbiological indicators from Traceo were based on selected strains of lactic acid bacteria. The color change from green to red occurred due to a pH change in a gel layer in which the bacteria grew (Suppakul 2012).

Applications of TopCryo™ labels include time–temperature monitoring of food products and beverages at cold chain temperatures from 2 to 12 °C and from a few hours up to 12 days. Some examples show the application of the label during packaging (Fig. 5.19), the condition of well-preserved meat (Fig. 5.20), and spoiled clams at a distributor (Fig. 5.21).

5.6 Smart Systems

Many smart solutions involve a combination of smart materials and other technological capabilities such as sensors, electronics, software, integrated circuits, or devices. Chromogenic materials applied on conventional support materials constitute *smart composites* with extraordinary performance as demonstrated by the projects and products illustrated so far. When smart composites are attached or integrated into control systems, they form smart or intelligent systems. *Smart systems* have the capability to sense environmental changes and to respond to them via an active control mechanism. The following case studies discuss applications which involve smart systems which offer more than the capabilities of smart materials or smart composites alone.

5.6.1 SymbiosisO, A Collection of Textile Interfaces by Kärt Ojavee and Eszter Ozsvald

SymbiosisO is the project of a collection of programmable textile interfaces with the capacity to visualize information and stimulate emotions. It is the result of collaboration between Estonian designer and researcher Kärt Ojavee and Hungarian designer, engineer, and multimedia artist Eszter Ozsvald based in New York (SymbiosisO 2013). The two designers have met during a period of work at the Estonian Biorobotics Center and in 2009 they have started collaboration for the development of applications with interactive materials, which generated the project SymbiosisO. This project, carried out in several steps, was presented in the form of realization-installation of an active skin with the objective of inserting the dynamic rhythms of nature in the artificial landscape of daily life. The project suggests the possibility of a "romantic" interaction between man and computer. The use of soft materials (fabrics, felt, and padding) and the dynamic behavior of thermochromic surfaces which react to the impulse of human touch or changes in environmental conditions contribute to this perceptive result. The animation that occurs in some of the prototypes produced at various steps of the design process consists of a gradual change of surface colors according to graphical patterns which are attributed to the effect of growth of natural organisms. In its complexity, the project has permitted to give life to various solutions prototyped and identified with the term *Symbiosis* with the addition of a different consonant or suffix: *Symbiosis C–S–W* and *:voxel*.

The first step of design research has focused on *SymbiosisW* (where *W* stands for Wall), a soft surface, slightly tridimensional, biomimetic because inspired by

---

[1] A second solution of *SymbiosisW* was realized by the connection of piezoelectric transducers to the electronic system in order to reduce friction.

Fig. 5.22 SymbiosisW,
2009. *Cortesy*
SymbiosisoO:voxelVoxel

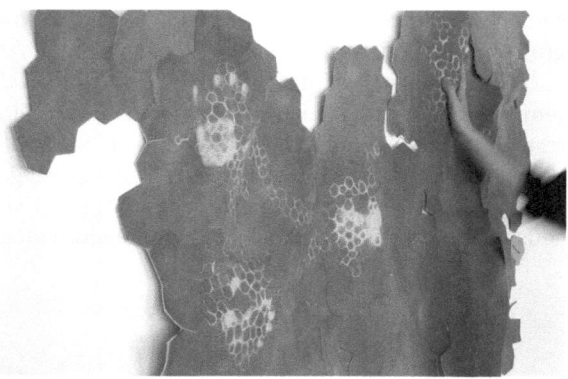

the structure of biological organisms, and consisting of a pattern of hexagonal cells (Fig. 5.22).

Thanks to an electronic system equipped with tactile capacitive sensing,[1] the surface detects human touch and responds with a mutation of the visual appearance. Through the reduction of color across a geometric network, this mutation simulates the growth of the smallest cells below the hand of the user based on the generative model that spreads on the surface with the effect of the passing hand. By the intention of the designers, the project wants to stimulate a reflection on the evolution of natural environment: not only does the human civilization have an impact on nature but also nature reacts and adapts to these changes. It is thus possible to describe the dynamic effect of *SymbiosisW* as the spread of a new species of lichen. The survival rate of lichen species suddenly drops as the air

Fig. 5.23 SymbiosisS.
Courtesy SymbiosisO:voxel

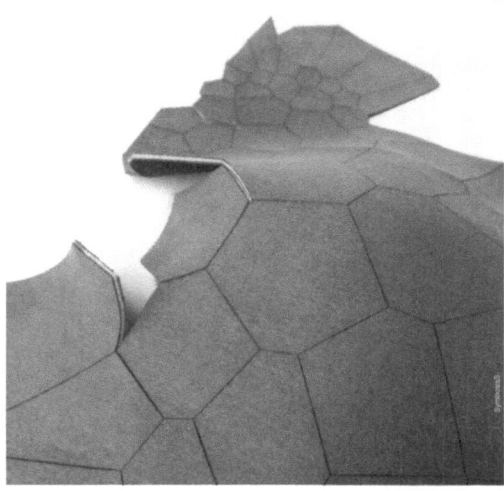

becomes polluted but new species have adapted to civilization and now they can grow on artificial surfaces.

The second step of the project led to *SymbiosisS* (where *S* stands for *seat*), a semi-horizontal surface in felt which greets the visitor, inviting her to sit down and rest on a soft, slightly three dimensional material because it adopts a surface structure characterized by Voronoi tessellation (Fig. 5.23). This generative model facilitates changes in superficial form through a slight sliding movement which adds dynamism to the surface along its major extension.

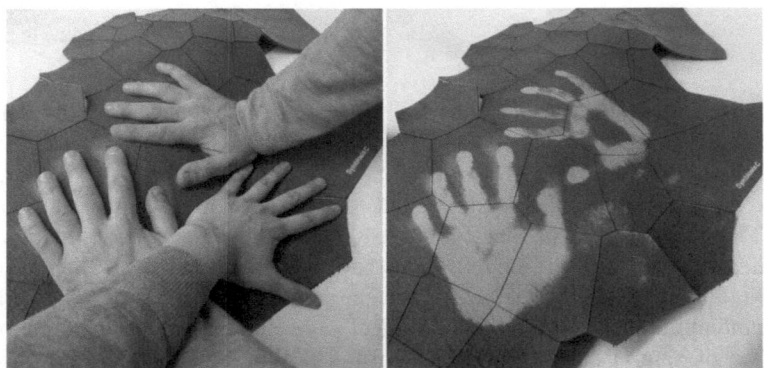

Fig. 5.24 SymbiosisC. *Courtesy* SymbiosisO:voxel

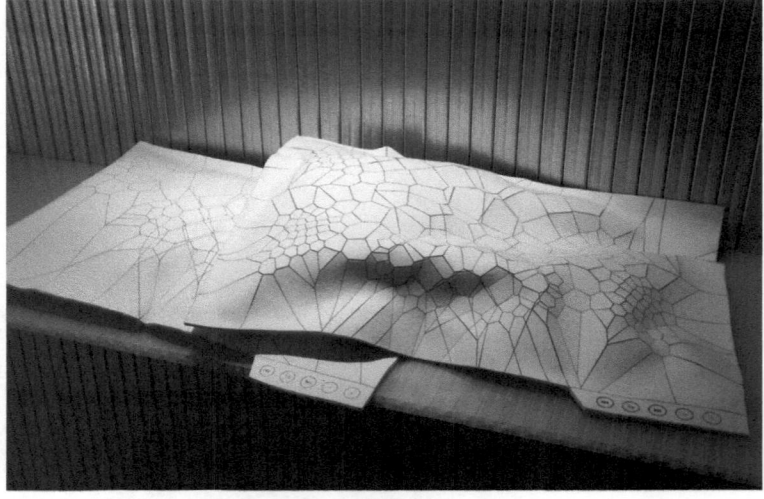

Fig. 5.25 Play Me! *Cortesy* Symbiosiso:voxel

When someone sits down, one of the soft switches hidden between the two layers of fabric excites the electronic control system that triggers an increase in temperature, leading to color change of the surface and simultaneous blocking of movement. Once the "disturbance" calms down, the model continues its slow expansion. This interaction between change of shape and color is one of the many possible combinations and gives an idea of the potentials of tangible textile interfaces. Each installation can be customized as desired. The felt has been chosen as surface material due to its good acoustic insulation, heat conservation, and biodegradability. According to the designers, *SymbiosisS* is an ideal solution for public or semi-public environments such as the waiting areas, which may be used to calm down stress, provide information, and engage the users. The application of intelligent *SymbiosisS* systems is a perfect alternative to the vibrant LCD screens which give commercial information while waiting.

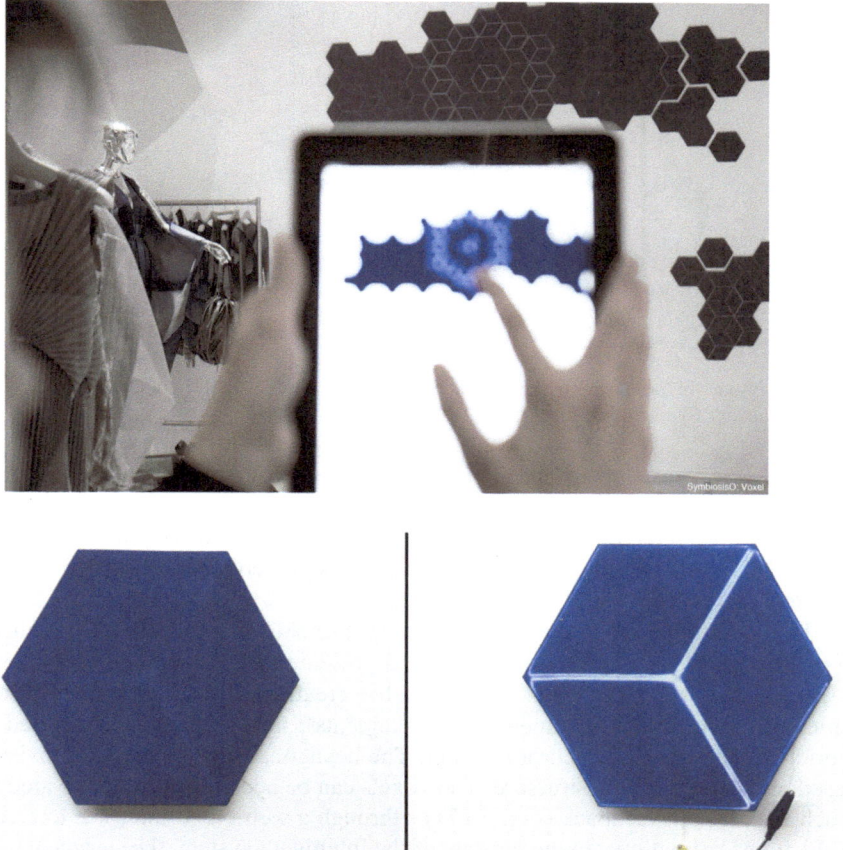

Fig. 5.26 SymbiosisO:voxel, interactive design installation at Tribeca Issey Miyake, April 2012 (*upper image*). Voxel in *off* and *on* position (*lower image*). *Courtesy* SymbiosisO:voxel

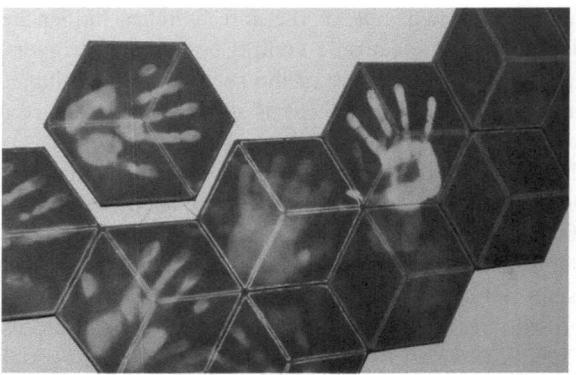

Fig. 5.27 Thermochromic ink on Voxel. *Cortesy* SymbiosisO:voxel

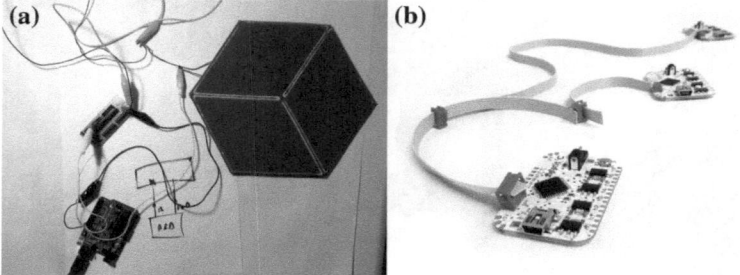

Fig. 5.28 Voxel in *on* position during experimentation (*left*), and Heatit°C microcontroller (*right*). *Courtesy* SymbiosisO:voxel

SymbiosisC (*C* stands for cushion) is the low-tech component of the collection; a cushion that responds to the heat of body and warm objects by changing its color (Fig. 5.24). From the combination of *SymbiosisC* and *SymbiosisS* came *Play Me!* (Fig. 5.25), the first product of the series made in collaboration with K–O–I Design.

The last installation of the collection is *SymbiosisO*: *voxel* (Fig. 5.26) realized in collaboration with artist Alex Dodge and presented at Tribeca Issey Miyake in 2012. *SymbiosisO*: *voxel* is a system for the creation of display surfaces, constructed by 64 voxels (volumetric pixel elements), hexagonal panels covered in fabric treated with thermochromic paint. The hexagonal shape allows a provision based on the honeycomb structure. The voxels can be activated and deactivated by touch with instant feedback (Fig. 5.27) or through a web-based interface to create multi-frame animations. In the *off* state of the intelligent system, the surface has an intense blue color; in the *on* state, a white graphic pattern appears (an edge that shows a partition inside voxels, depicting a cube seen in perspective view).

For the activation of thermochromic ink, *Heatit°C* (Fig. 5.28), a microcontroller based on open-source electronics platform, is used. This provides a precisely controlled high current which contains the transistors and controls the sensors. To heat the fabric along the desired pattern, the system uses individually controlled resistive nickel-chrome wires.

With *SymbiosisO: voxel*, the designers' intention is to give the system the advantages of modularity and flexibility: the surface does not have a maximum dimension or a predetermined scheme, it does not imitate biological models of growth, but is made up of modular elements (voxels) that are connected to each other to generate the interface. A single *voxel* is a flexible display equipped with software which allows the users to create their own personal graphics and animation. Compared to other solutions, the system represents a step forward towards mass production but also to individual production. In fact, to produce *SymbiosisO: voxel* the designers supply a CAD program for digital manufacturing.

5.6.2 Fabrication Bag by Hanna Landin and Linda Worbin

Linda Worbin, researcher in Textiles and Interaction Design at the Swedish School of Textiles is exploring smart materials for some time to create interactive textile

Fig. 5.29 Fabrication Bag, 2005. *Courtesy* L. Worbin

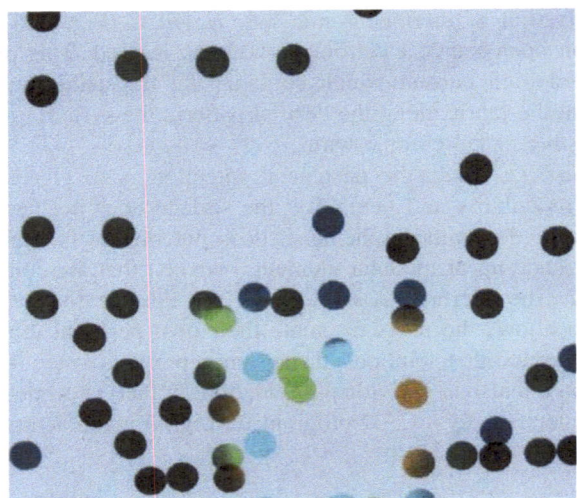

Fig. 5.30 Textile with *dots* in thermochromic ink for Fabrication Bag, 2005. *Courtesy* L. Worbin

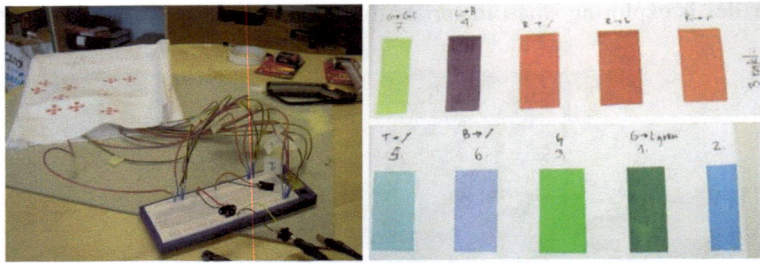

Fig. 5.31 Fabrication Bag during the experimentation *(left)* and color palette of thermochromic ink (Variotherm 27 °C) *(right)*. *Courtesy* L. Worbin

(Worbin 2010, Nilsson et al. 2011). The Fabrication Bag may be defined as an investigation on a pattern of static forms that change color, or an investigation on a dynamic textile pattern of static form, by digital information (mobile phone data). In this project, thermochromic ink was applied on textiles in cotton combined with conductive yarn on the backside to control the color change of the thermochromic print.

The concept of the Fabrication Bag (Fig. 5.29) is meant to be an accessory for mobile phones that replaces the sound and vibration signals with changing colors in textile patterns on the outside and inside of a bag. Initially the fabric of the bag is white with grey dots in different nuances, and when the phone receives calls or text messages the colored dots slowly change from dull to colorful.

The project uses thermochromic heat sensitive textile pigments (Fig. 5.30) that change from grey to five different colors: pink, light blue, light green, yellow, and

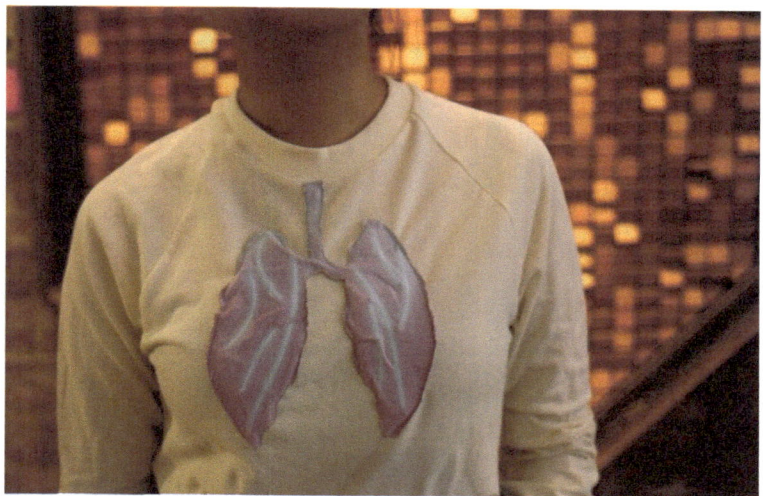

Fig. 5.32 Pollution monitoring fashion by Sue Ngo and Nien Lan, 2011

orange. The smart system was completed by a 14 V battery for power supply, a program basic X and a microcontroller BX24.

The textile of the bag contains multiple layers of thermochromic ink printed in dots that are activated by heat patches underneath (Fig. 5.31). Nine heating elements are mounted inside the bag and when a heating element is turned on (individually or combined in group), the surface print will change color indifferent of expressions. When an incoming call is detected, different areas of the heat patches are activated for localized heating of the thermochromic print. The idea behind the programming was to connect different kinds of data, calls etc. with different visual design expressions. Depending on the level of code abstraction, the bag will be more or less easy to "read".

The relation between data and visual expression is able to introduce new design possibilities and variables. The data contributes to the visual expression of dynamic pattern, varying from time to time. It can be done in real time, or recorded in advance or with a delay. The project also suggests and exemplifies new behaviors of textile patterns as well as interaction with mobile phones.

5.6.3 Pollution Monitoring Fashion by Sue Ngo and Nien Lan

Sue Ngo e Nien Lan, New York City based interaction designer and programmer, respectively, designed and prototyped a series of pollution monitoring sweatshirts during their Masters in Interactive Telecommunications Program at New York University. Drawing inspiration from the hypercolor T-shirts of yore, the *Warning*

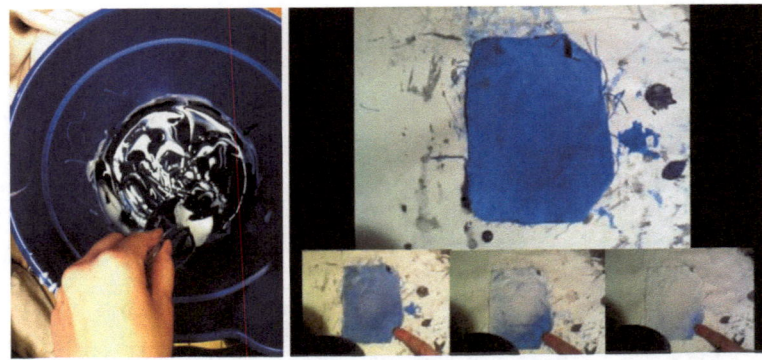

Fig. 5.33 Sue Ngo mixing thermochromic ink (*left*), and testing the color results (*right*). *Courtesy* S. Ngo

Signs line was born. These garments are capable of emitting visible signals of air pollution, usually in the form of invisible organic volatiles (Huffington Post 2011).

Among the prototypes created by the two young designers, there are sweatshirts which are white and bear a pink heart or a set of lungs in the front made of thermochromic fabric. The shirts emit a warning sign when in contact with high levels of carbon monoxide (CO): the veins on the lungs or heart, firstly invisible, subtly change color from a healthy pink to a slightly worrying blue-grey, as if to indicate that the CO is penetrating the human body and reaching the organs (Fig. 5.32).

The transformation of the sweatshirt is made possible by a smart system hidden between the two layers of textile in the front of the T-shirt, composed of a gas sensor, MQ-7, thermochromic ink (Fig. 5.33), a resistive wire for heating, connector wires, and a powerful micro-controller. The MQ-7 sensor reveals the concentration of CO in the air nearby, thanks to a semiconductor layer of gas sensor made of tin dioxide (SnO_2). This sensor detects the presence of gas in the

Fig. 5.34 Photochromic sculpture made visible by UV radiation. *Courtesy* Tomoko HashidaHashida

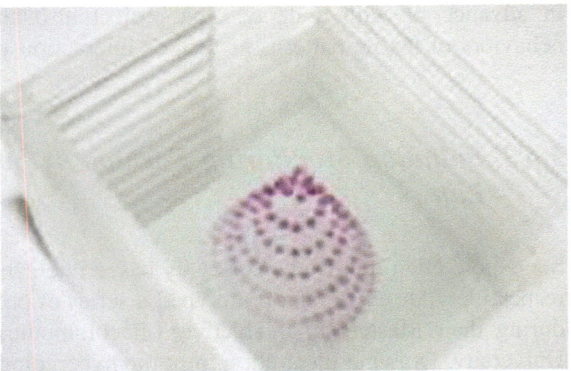

range of 20–2,000 parts per million (ppm), with very rapid response times, and provides an output as resistivity directly proportional to the gas concentration.

5.6.4 Photochromic Sculpture by Tomoko Hashida, Yasuaki Kakehi, and Takeshi Naemura

This is a project realized by a team from the University of Tokyo whose members are Tomoko Hashida, Yasuaki Kakehi, and Takeshi Naemura (Hashida et al. 2011a, b). The system which creates a "photochromic sculpture" consists of mainly two parts. The control part, which is basically a projector, provides the ultraviolet (UV) light source when desired. The control part is a hidden system which contains a UV light source at 365 nm and a digital micromirror that is able to control two-dimensional invisible patterns dynamically. The second part is a three-dimensional display which becomes visible under UV light. A photochromic sculpture is created by several layers of transparent plates coated with photo-chromic spiropyran granules. Spiropyrans are a group of organic photochromic molecules which have reversible switching capabilities. Their coloration-bleaching cycle can be repeated up to 1,000 times (Urban 2011; Charvat 2004). When UV radiation hits the stacked layers, colored pixels appear within seconds (Fig. 5.34). Similarly, when the UV radiation is blocked, the colors gradually fade away until the layers become transparent. One requirement in this system is that granules in upper layers should not block those underneath so that the UV light can effectively

Fig. 5.35 Ron Arad, No Bad Colour workstation applying Active True Colour, Salone del Mobile, Milan 2013

hit the desired pixels in the lower layers. The sculpture made of colorful pixels can be dynamically modified by changing the pattern of UV light from the control part.

Photochromic sculpture can be regarded as both art and design. The system is suitable for innovative displays and advertisement both small and large. The project team is developing a UV projector for more complicated patterns and for larger scale photochromic sculptures for outdoor applications.

5.6.5 No Bad Colours by Ron Arad, with Active True Colour

For the occasion of Salone del Mobile 2013, the famous designer Ron Arad, along with the company Versatile, presented the first *Active True Colour* piece, an integrated color-changing workstation containing a desk, shelves and wall, within his new project, *No Bad Colours* as part of *Office For Living*, a Jean Nouvel curated exhibition (Fig. 5.35).

The uniqueness of the workstation by Ron Arad lies in the application of the Active True Colour technology which is composed of interconnected *tiles* which form a dynamic surface that enables infinite and instant changes in colour. In the solution *No Bad Colours* the tiles size was 15×15 cm in green, red, yellow and black colors but other color ranges are available. Each one was activated individually allowing large walls to be made showing many colors and thus an infinite range of colored patterns is possible.

Unlike other technologies like LED, LCD or Plasma, already well known and applied, colors emitted from the surface are reflecting rather than transmitting, so the solution does not require backlit panels or energy intensive light-emission and the surfaces enjoy rather than suffer from external light. Thus its brightness is always the same relative to the ambient lighting so it provides a non obtrusive background color for walls and other large areas.

The device is also bistable—once a state (of any color) is driven to it will remain in that state until told to change to another state. Thus power is only used to change the state of the material and not to maintain that state—again this leads to very low power use. The device in Milan was using 20 W total (about 5 W/m^2).

The dynamic play of colors created by this new technology, which can also be shaped into patterns realized with vinyl masks or through milling of glass surfaces, is able to enrich products and spaces (living, public, and workspaces) by expanding the way people experience the environments and time, overcoming the dullness of certain solutions, thanks to the management of time and colors through a dedicated software.

Active True Colour, the technology which makes such dynamic solutions possible, is patented by Versatile, a technology driven company. The inventor is David Coates, Chief Technology Officer of Versatile, who applied his knowledge in organic chemistry and used an electrochromic fluid placed between three transparent plastic layers. This special proprietary liquid crystal is capable of selectively reflecting a range of colors. Upon the application of a slight energy

Fig. 5.36 Wall in Active True Colour with graphic pattern

impulse, the fluid molecules change orientation, reflecting the light in a different way such that a different color is generated at the surface. This sandwiched liquid crystal can be optimized to show one of many colors. This color can be electrically driven to show several grey levels of the same basic color. Other colors can be added by using other layers of liquid crystal (usually 1–3 layers are used) (Coates 2012; Versatile 2013).

The effect of color modulation is repeated according to the programmed electric impulses. The film can be switched electrically between a transparent state and a reflective colored state. The software to drive the tiles is proprietary. Each tile can be driven individually and thus show any of the colors available to it together with grey levels. The colors can also be gradually faded from the full color state to the black state.

The activation of the device with a dynamic change in color appears to be generated by a light source but in reality there is no light emission and the device does not need any artificial light source. The surface does not require any electric current in order to maintain the generated color. Electricity is only needed to modify the color. A minimal amount of electricity is consumed by the device during its function.

Active True Colour is a technology with great potential for design and architecture with an adaptive palette of color, patterns and architectural finishes (Fig. 5.36). Its use will permit to renew furniture or to modify the walls of

Fig. 5.37 Footstool with a geometric dynamic pattern, 2011. Photo of Jan Berg, *courtesy* L. Worbin

domestic and public environments. It will be possible for example to adopt interiors to themes of various seasons, to the mood of people who live in them, or to different activities being performed during the day or the week.

5.6.6 Footstool with a Geometric Dynamic Pattern by Linnéa Nilsson, Mika Satomi, Anna Vallgårda and Linda Worbin

In this project, the team composed of Linnéa Nilsson, Mika Satomi, Anna Vallgårda and Linda Worbin from University of Borås, Sweden, collaborated with the firm Ire Möbel in order to investigate possible applications of thermochromic fabric in the production of furniture with interactive capacities.

The project, exhibited at the Stockholm Furniture Fair and at Salone del Mobile in Milan, experiments with a particular structure of upholstery on a footstool

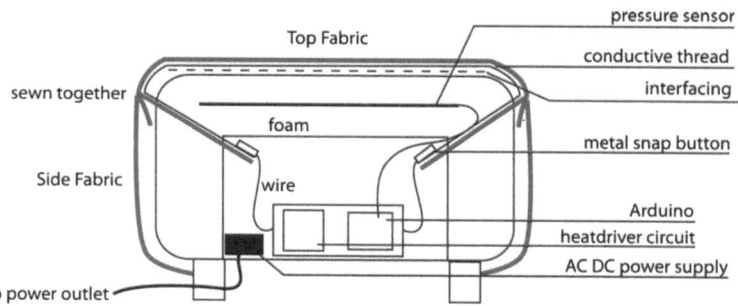

Fig. 5.38 Section of the footstool showing the electrical and electronic components placed inside the stool. *Courtesy* L. Worbin

(Fig. 5.37), equipped with a smart system which controls the dynamism of patterns printed on the fabric and interacts with the user who sits on the stool.

The upholstery uses a cotton fabric with geometric patterns printed using conventional pigments and thermochromic ink, which at 27 °C change from opaque to transparent. Sewn on the back of the fabric, there are steel wires used as conductors, placed every 5 mm in the weft direction. The smart system is completed by two pressure sensors, positioned below the fabric, which turn on the power as soon as someone sits on the stool, *Arduino* boards, placed inside the footstool, which control the current which circulates in different conductive wires, circuits and elements of feed and control (Fig. 5.38). When the electric current, driven by pressure sensors, passes through the wires, these start to warm up and transmit heat to the fabric, promoting color changes in the pattern along each wire.

The experiment aims to study expressive possibilities of the fabric with dynamic patterns generated thanks to the use of thermochromic inks managed by a

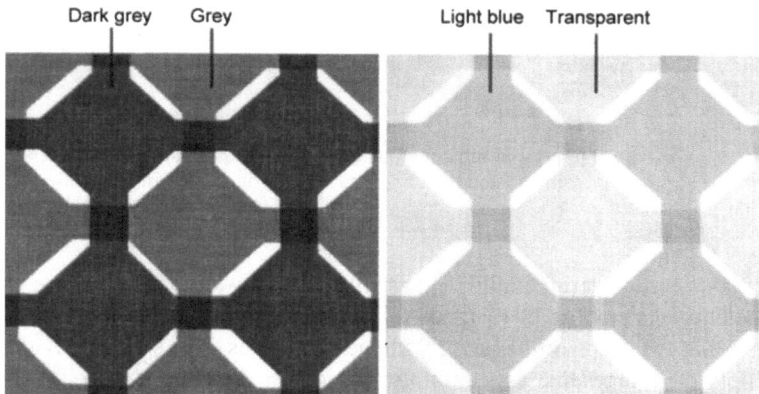

Fig. 5.39 Detail of print on the fabric at a temperature <27 °C (*left*) and >27 °C (*right*) *Courtesy* L. Worbin

Fig. 5.40 *Top row* detail of print in full color palette at <27 °C (*left*) and >27 °C (*right*); *middle* and *bottom rows*: same pattern geometry with different colors during thermal transition. *Courtesy* L. Worbin

smart system, considering multiple variables in play (the shape of the pattern, color palette, and rhythm of dynamism) and their interference.

Regarding the shape of the pattern, the geometry has been specifically designed to multiply the images that appear on the fabric. The designed pattern is actually a composition of different patterns that are revealed in time. In fact, some parts of the geometry are realized with conventional colorants and others with

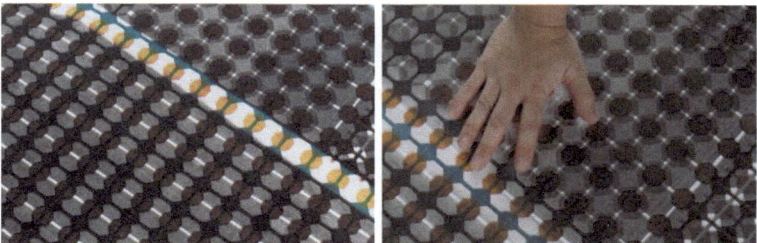

Fig. 5.41 Some of the expressions of pattern that are generated at different temperatures to which the fabric is exposed to. Photo of Jan Berg, *courtesy* L. Worbin

thermochromic inks. When the thermochromic ink becomes transparent, the shape of the pattern changes (interference of the shape with the color palette). Furthermore, stimulating thermochromic inks with slightly different transition temperatures (interference of the shape with time variable) it is possible to obtain variations of color at different phases with respect to each other, which is expressed as new variations of the geometric pattern.

Regarding the colors of the pattern, the combination of a conventional and a thermochromic ink during printing on parts of geometric surfaces that overlap with diverse shapes (color/shape interference) allows obtaining more colors during the transition, instead of a simple switch from one color to another. For example, it is possible to switch from dark gray to blue or green to yellow using colors that are composed by addition or subtraction of thermochromic colors. Moreover, each

Fig. 5.42 Footstoll with a geometric dynamic pattern, 2011. Photo of Jan Berg, *courtesy* L. Worbin

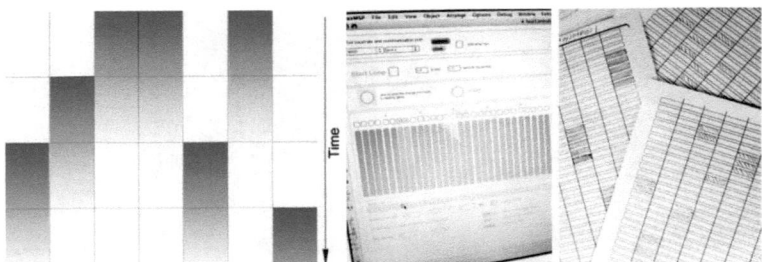

Fig. 5.43 Program sheet for the timing of resistive heating wires (*left*) and the interface adopted for the programming of the general behavior of the footstool. *Courtesy* L. Worbin

new color added to the base palette introduces a whole series of shade that combine with other colors (Figs. 5.39, 5.40, 5.41). The addition of a small amount of thermochromic magenta pigment to thermochromic gray determines changes of color from gray to transparent passing through different shades of magenta. The interference between the color combinations during the transition with different pattern transition times, multiplies the nuances which appear simultaneously on the object, generating a complete range of colors.

Regarding the rhythm of pattern, the time dimension not only influences the design of color variables and shape but also the specific design of a time-shape, the rhythm. This variable is affected by the time required to reach the transition temperature, which ranges from 10 to 20 s, while the cooling may require several minutes. During this time, the pattern on the surface changes, displaying the diverse expressions of pattern and intermediate shades that produce a significant overall visual appearance. The time/rhythm variable may be controlled by programming the heating time of the different conductive wires below the fabric, with the Arduino board. A gradual temperature variation between a wire and another allows the phase shift of color transitions, resulting one to appreciate the graduality of the change of different colors and shades that are generated during transition at different moments in different parts of the footstool surface.

Depending on the time of heating, certain parts of the upholstery surface may be in their original state, in a heated stated, or in a state of gradual change (Fig. 5.42).

In addition to all this, there is also the interaction with the user. In this case, the temporal expressions could overlap and thus create new, unexpected combinations. Having to handle all these variables (forms of pattern, color palette, time/rhythm) and their inference, the general interface design is quite challenging. It is equivalent of deconstructing different elements of the composition in order to design diverse compositions (layout) and then a composition of compositions. The composition of each layout is determined by position, size, and intensity of the heated zones. The number of possible combinations is very large but it is necessary to verify the changes and transitions of patterns. This operation could result in the elimination of some of the combinations. Due to its complexity, the design process could become very difficult without the appropriate tools. The team has used the tools which are typically used to design textile patterns (sketches, CAD programs, prototypes etc.) but they also developed a different instrument: a graphic interface for programming the pattern which is similar to a sheet of music (Fig. 5.43), a combination of "scores" where it was possible to mark the heating of specific sections and keep track of previous and consecutive layouts.

With this experimental work, the team has determined a series of practices and complexities which emerge when chromogenic materials are used in smart systems. The complexity could be transformed into possibilities of expression if mastered properly. The question is thus how to master the complexity.

References

Abrahamson M (2009) Absent Traces: Jurgen Mayer and the saturated surface, Blog. http://criticundertheinfluence.wordpress.com/2009/10/13. Accessed 11 Sept 2012

Charvat RA (ed) (2004) Coloring of plastics. Wiley, Hoboken

Coates D (2012) Cholesteric reflective display. In: Chen J, Cranton W, Fihn M (eds) Handbook of visual display technology. Springer, Berlin

Cory J (2007) Realising the endless: the work of Jürgen Mayer H. and the legacy of Frederick Kiesler. Papers Surrealism 5(Spring):1–18

Erickson B (2009) Self-darkening eyeglasses. Sci Technol 87(15):54

Foundation for the Globe of Science and Innovation (2012). http://www.fondationglobe.ch/index.php?id=2&L=1. Accessed 25 Sept 2012

Granqvist CG, Green S, Niklasson GA, Mlyuka NR, von Kraemer S, Georen P (2010) Advances in chromogenic materials and devices. Thin Solid Films 518:3046–3053

Hashida T, Kakehi Y, Naemura T (2011) Photochromic sculpture: volumetric color-forming pixels. SIGGRAPH 2011, Vancouver, British Columbia, Canada, 7–11 Aug 2011

Hashida T, Kakehi Y, Naemura T (2011) Photochromic sculpture. http://nae-lab.org/~hashida/pSculpture.html. Accessed 1 Aug 2012

Huffington Post (2011) T-Shirt detects pollution: 'Warning Signs' by Nien Lam, Sue Ngo. http://www.huffingtonpost.com/2011/01/25/tshirt-measure-pollution-_n_813356.html. Accessed 10 July 2013

Madison Gas and Electric Co. (2012). https://www.mge.com/business/saving/madison/PA_6.html. Accessed 26 Sept 2012

Mayer JH (2005) In Heat, description of the installation by the architect, J. Mayer H. Architekten. http://www.jmayerh.de. Accessed 12 Sept 2012

Nilsson L, Vallgårda A, Worbin L (2011) Designing with smart textiles: a new research program. Nordic Design Research Conference 2011, Helsinki. www.nordes.org

Oriakhi CO (2009) Intelligent chemical indicators. In: Schwartz M (ed) Smart materials. CRC Press, Boca Raton

Potter L, Campbell A, Cava D (2008) Active and intelligent packaging—a review. CCFRA, Gloucestershire

Rainbow Winters (2012) Catalog, Spring-Summer 2011. www.rainbowwinters.com. Accessed 17 Aug 2012

Smart Glass International (2011) Electronically switchable glass handbook. http://www.smartglassinternational.com/downloads/LC_SmartGlass_Handbook.pdf. Accessed 26 Sept 2012

Suppakul P (2012) Intelligent packaging. In: Sun DW (ed) Handbook of frozen food processing and packaging. CRC Press, Boca Raton

SymbiosisO (2013) SymbiosisO collection of textile interfaces. http://www.symbiosiso.com. Accessed 11 July 2013

Urban MW (ed) (2011) Handbook of stimuli responsive materials. Wiley, Weinheim

Van Gemert B (2000) The commercialization of plastic photochromic lenses: a tribute to John Crano. Molecular Crystals and Liquid Crystals Science and Technology. Section A. Mol Cryst Liq Cryst 344(1):57–62

Versatile (2013) Versatile forever changing colour. http://www.foreverchangingcolour.com. Accessed 10 July 2013

Voermanek K (2012) J. Mayer H. Architekten, Apple talk. http://www.baunetz.de/talk/crystal. Accessed 11 Sept 2012

Worbin L (2010) Designing dynamic textile patterns. Dissertation, Chalmers University of Technology, Sweden

Younger Optics (2012) Drivewear catalog. http://www.drivewearlens.com. Accessed 18 Sept 2012

Index

M. Ferrara and M. Bengisu, *Materials that Change Color*, PoliMI SpringerBriefs,   137
DOI: 10.1007/978-3-319-00290-3, © The Author(s) 2014